T0191578

Studies in Computational Intelligence

Volume 666

Series editor

Janusz Kacprzyk, Polish Academy of Sciences, Warsaw, Poland
e-mail: kacprzyk@ibspan.waw.pl

About this Series

The series "Studies in Computational Intelligence" (SCI) publishes new developments and advances in the various areas of computational intelligence—quickly and with a high quality. The intent is to cover the theory, applications, and design methods of computational intelligence, as embedded in the fields of engineering, computer science, physics and life sciences, as well as the methodologies behind them. The series contains monographs, lecture notes and edited volumes in computational intelligence spanning the areas of neural networks, connectionist systems, genetic algorithms, evolutionary computation, artificial intelligence, cellular automata, self-organizing systems, soft computing, fuzzy systems, and hybrid intelligent systems. Of particular value to both the contributors and the readership are the short publication timeframe and the worldwide distribution, which enable both wide and rapid dissemination of research output.

More information about this series at http://www.springer.com/series/7092

Michael Mutingi · Charles Mbohwa

Grouping Genetic Algorithms

Advances and Applications

 Springer

Michael Mutingi
Faculty of Engineering
Namibia University of Science and
  Technology
Windhoek
Namibia

and

Faculty of Engineering and the Built
  Environment
University of Johannesburg
Johannesburg
South Africa

Charles Mbohwa
Faculty of Engineering and the Built
  Environment
University of Johannesburg
Johannesburg
South Africa

ISSN 1860-949X   ISSN 1860-9503 (electronic)
Studies in Computational Intelligence
ISBN 978-3-319-83048-3   ISBN 978-3-319-44394-2   (eBook)
DOI 10.1007/978-3-319-44394-2

Printed on acid-free paper

This Springer imprint is published by Springer Nature
The registered company is Springer International Publishing AG
The registered company address is: Gewerbestrasse 11, 6330 Cham, Switzerland

This book is dedicated to operations analysts, computational scientists, decision analysts, and industrial engineers

Preface

Recent research trends have shown that industry is inundated with grouping problems that require efficient computational algorithms for grouping system entities based on specific guiding criteria. Grouping problems commonplace in industry include vehicle routing, container loading, equal piles problem, machine-part cell formation, cutting stock problem, job shop scheduling, assembly line balancing, and task assignment. These problems have a group structure with identifiable characteristic features, that is, the need to form efficient groups of entities according to guiding criteria, and the need to allocate those groups to specific assignees in order to satisfy the desired objectives. It is interesting to note that, across all the spectrum of these problems, grouping and allocation criteria are inherently very similar in nature.

The wide spectrum of real-world grouping problems, the striking similarities between their features, and the multi-criteria decisions involved are three major motivating factors behind the research momentum in this area. However, more challenging issues in this field have appeared in recent researches. First, there is an ever-growing need to address uncertainties in various grouping problem situations. Second, decision analysts in the field often call for multi-criteria decision approaches by which multiple criteria can be handled simultaneously. Third, researchers and decision analysts have realized the need for interactive, population-based algorithms that can provide alternative solutions rather than prescribe a single solution to the decision maker. Examples of such approaches are tabu search, particle swarm optimization, ant colony optimization, simulated evolution algorithm, simulated metamorphosis algorithm, genetic algorithms, and grouping genetic algorithms. Thus, in sum, recent research has emphasized the need for development of interactive multi-criteria computational algorithms that can address grouping problems, even in uncertain or fuzzy environments.

Evidently, notable research has focused on advances in genetic algorithms and related hybrid approaches, with application in various problem areas. Current research trends tend to show that there is a high potential for remarkable advances in genetic algorithms and its variants, specifically in grouping genetic algorithms. Genetic algorithm-based approaches offer a more user-friendly, flexible, and adaptable population-based approach than related algorithms. Given these advantages, further developments and advances in grouping genetic algorithms are quite promising.

The purpose of this book is to provide an account of recent research advances and, above all, applications of grouping genetic algorithm and its variants. The prospective audience of the book "Grouping Genetic Algorithms: Advances and Applications" includes research students, academicians, researchers, decision analysts, software developers, and scientists. It is hoped that, by going through this book, readers will obtain an in-depth understanding of the novel unique features of the algorithm and apply it to specific areas of concern.

The book comprises three parts. Part I presents an in-depth reader-friendly exposition of a wide range of practical grouping problems, and the emerging challenges often experienced in the decision process. Part II presents recent novel developments in grouping genetic algorithms, demonstrating new techniques and unique grouping genetic operators that can handle complex multi-criteria decision problems. Part III focuses on computational applications of grouping genetic algorithms across a wide range of real-world grouping problems, including fleet size and mix vehicle routing, heterogeneous vehicle routing, container loading, machine-part cell formation, cutting stock problem, job shop scheduling, assembly line balancing, task assignment, and other group technology applications. Finally, Part IV provides concluding remarks and suggests further research extensions.

Johannesburg, South Africa   Michael Mutingi
                                                                      Charles Mbohwa

Contents

Part I
Introduction

Chapter 1
Exploring Grouping Problems in Industry

1.1 Introduction

In many real-world industry settings, it is often desirable to improve the performance of systems, processes, and products by partitioning items into groups, based on suitable decision criteria. For instance, in logistics management, decision makers wish to minimize the overall transportation costs, the number of vehicles used, and the average waiting time experienced by the customer (Taillard 1999; Potvin and Bengio 1996). To achieve this, it is important to ensure that customers are portioned into efficient groups to be visited by a set of vehicles. As such, it is important to optimize grouping of customers to be visited by each vehicle, while considering the size, type, and capacity of the available vehicles. Similarly, when assigning a set of tasks to a team of workers, it is crucial to form cost-effective and efficient groups of tasks that can be assigned to workers in an optimal manner. Furthermore, manufacturers may want to find the best way to group parts with similar characteristics so that similar parts can be produced using specific processes in specific departments. Some other well-known problems include bin packing problem, load balancing (or equal piles) problem, and machine cell formation problem in group technology. The major task in these problem situations is to group (or partition or cluster) a set of items into disjointed subsets or groups, in some optimal or near-optimal manner. Such problems are commonplace across a wide range of industries, from manufacturing to service industry. In this book, these problem situations are called *grouping problems*.

It is important to note at this point that grouping problems are known to be combinatorial, hard and computationally difficult to solve (Taillard 1999; Potvin and Bengio 1996; Moghadam and Seyedhosseini 2010). On the other hand, it is also important to realize that grouping problems have common grouping features and characteristics that can be potentially be exploited in order to develop more effective solution approaches. In a nutshell, further studies on the various types of grouping problems in the literature revealed interesting facts as outlined below:

© Springer International Publishing Switzerland 2017
M. Mutingi and C. Mbohwa, *Grouping Genetic Algorithms*,
Studies in Computational Intelligence 666,
DOI 10.1007/978-3-319-44394-2_1

1. The optimization of grouping problems is usually defined in terms of the composition of the groups of items and the overall array of all the groups;
2. Grouping problems possess a grouping structure that can be utilized for developing effective computational algorithms;
3. Grouping problems are highly combinatorial in nature, NP-hard, and computationally expensive;
4. Grouping problems are highly constrained, which adds to their computational complexity; and
5. Some of the variables of grouping problems are not precise, so much so that fuzzy modeling is a useful option.

Due to their computational complexity, the use of heuristics, expert systems, and metaheuristic approaches are a viable option for solving various grouping problems. Examples of these approaches are tabu search, particle swarm optimization (Mutingi and Mbohwa 2014a, b, c), genetic algorithms, grouping genetic algorithm (Mutingi and Mbohwa 2013a, b), and other evolutionary algorithms.

Genetic algorithm (GA) is a potential solution approach for this class of problems (Rochat and Taillard 1995; Badeau et al. 1997). GA is a metaheuristic approach based on the philosophy of genetics and natural selection. In its operation, GA encodes candidate solutions into chromosomes (or strings) and improves the strings by copying strings according to their objective function values and swapping partial chromosomes to generate successive solutions that improve over time. Its distinctive feature is the use of probabilistic genetic operators as tools to guide the search toward regions of the search space with likely improvement. Grouping genetic algorithm (GGA), originally developed by Falkenauer (1992), is a modification of the conventional genetic algorithms for addressing grouping problems. Some recent remarkable improvements and applications of the GGA exist in the literature (Mutingi et al. 2012; Mutingi 2013; Mutingi and Mbohwa 2013a, b).

Given the complexities of grouping problems, and their widespread occurrences in real-world industry, developing flexible, efficient, and effective solution methods is vital. The purpose of this chapter is to explore and identify grouping problems and explain their grouping characteristics. In this vein, the learning outcomes for the chapter are as follows:

1. To be able to identify various types of grouping problems and their common characteristics;
2. To develop an understanding of how to exploit grouping features of grouping problems for effective modeling; and
3. To understand the various modeling approaches for grouping problems and to open up research avenues for more efficient hybrid approaches.

The rest of this chapter is organized as follows: The next section identifies typical grouping problems in industry, their group structures, and shows how the problems lend themselves to grouping algorithms. Section 1.3 outlines past modeling approaches for grouping problems. Finally, concluding remarks and further research prospects are presented in Sect. 1.4.

1.2 Identifying Grouping Problems in Industry

In this section, a number of grouping problems from a wide range of industry settings are explored and identified, briefly illustrating their group structure and how they lend themselves to grouping genetic algorithm approaches. It was observed in this study that many of these problems loosely fall into subcategories such as manufacturing systems, logistics and supply chain, healthcare services, design, and other services. Some of them, such as team formation, timetabling, and economics, frequently occur across several types of industries. Table 1.1 provides a summary of these problems, together with selected references for further reading.

It is interesting to note that grouping problems are prevalent in a wide range of industry types. Several, if not all of these, problems possess similar characteristics and, therefore, lend themselves to a common group modeling and solution approach. By taking advantage of the knowledge of the common characteristics of the problems, a flexible computational algorithm can be developed for solving the problems. Such a computational algorithm is expected to be flexible and robust enough to be adapted to a wide range of problems with little or no fine-tuning. Apart from the ease of adaptation to problem situations, the algorithm is expected to be able to solve large-scale industrial problems within a reasonable time frame. In most industry settings, such as logistics, decision makers may need to make decisions on real time, or at least within a short space of time, so much so that efficient and flexible decision support is extremely crucial. In view of these and other related reasons, developing an efficient, flexible, and adaptable grouping algorithm is imperative and significant.

1.2.1 Cell Formation in Manufacturing Systems

Cellular manufacturing is a lean system of making groups of products, each group with products that are similar in shape, size, and processing characteristics in a cell. A cell defines a group of team members, workstations, or equipment that are grouped together to facilitate operations by eliminating setup costs between operations. Cell formation has a direct positive impact on the planning activities of a manufacturing system and is aimed at improving efficiency and productivity of the manufacturing system (Filho and Tiberti 2006; Mutingi et al. 2012; Mutingi and Onwubolu 2012). Machines are grouped together into efficient clusters, each operating on a product family with little or no inter-cell movement of the products.

Figure 1.1 provides an example of a cellular manufacturing system and its group representation. Assume that the system is one of the solutions to cell formation problem. Part (a) indicates that the manufacturing system consists of 3 cells, that is, cell 1, cell 2, and cell 3, where each cell comprises machine groups (1,3,4), (2,6), and (5,7,8), respectively. Part (b) shows the group structure or group representation

Table 1.1 Identified grouping problems in industry

No.	Grouping problems	Selected references
1	Assembly line balancing	Rubinovitz and Levitin (1995), Sabuncuoglu et al. (2000a, b)
2	Bin packing	Falkenauer (1996), Kaaouache and Bouamama (2015)
3	Job shop scheduling	Chen et al. (2012), Phanden et al. (2012), Luh et al. (1998)
4	Cell formation	De Lit et al. (2000), Onwubolu and Mutingi (2001), Filho and Tiberti (2006)
5	Container loading	Althaus et al. (2007), Joung and Noh (2014)
6	Heterogeneous fixed fleet Vehicle Routing	Gendreau et. (1999a, b), Tutuncu (2010), Tarantilis et al. (2003, 2004)
7	Cutting stock/material cutting	Onwubolu and Mutingi (2003), Rostom et al. (2014), Hung et al. (2003)
8	Fleet size and Mix Vehicle Routing	Liu et al. (2009), Brandao (2008), Renaud and Boctor (2002)
9	Equal piles problem	Falkenauer (1995), Rekiek et al. (1999)
10	Group maintenance planning	Li et al. (2011), Van Do et al. (2013), Gunn and Diallo (2015), De Jonge et al. (2016)
11	Handicapped person transportation	Rekiek et al. (2006)
12	Task assignment	Cheng et al. (2007), Mutingi and Mbohwa (2014a, b, c), Tarokh et al. (2011)
13	Home healthcare scheduling	Mutingi and Mbohwa (2014a, b, c)
14	Multiple traveling salesperson	Kivelevitch and Cohen (2013), Carter and Ragsdale (2006), Bektas (2006)
15	Modular product design	Yu et al. (2011), Kreng and Lee (2004), Chen and Martinez (2012)
16	Order batching	Henn and Wäscher (2012), Henn (2012)
17	Pickup and delivery	Parragh et al. (2008), Chen (2013), Wang and Chen (2013)
18	Site/facility location	Pitaksringkarn and Taylor (2005a, b), Yanik et al. (2016)
19	Wi-fi network deployment	Landa-Toress et al. (2013), Agustın-Blas et al. (2011a, b)
20	Student/learners grouping	Chen et al. (2012a, b), Wessner and Pfister (2001), Baker and Benn (2001)
21	Team formation	Wi et al. (2009), Strnad and Guid (2010), Dereli et al. (2007)
22	Timetabling	Dereli et al. (2007), Pillay and Banzhaf (2010), Rakesh et al. (2014)
23	Reviewer group construction	Chen et al. (2011), Hettich and Pazzani (2006)
24	Estimating discretionary accruals	Höglund (2013), Back et al. (1996), Bartov et al. (2000)
25	Economies of scale	Falkenauer (1993, 1994)
26	Customer Grouping	Sheu (2007), Chan (2008), Ho et al. (2012)

of the system. By considering process flows and the parts to be manufactured, multiple manufacturing system configurations can be generated and evaluated.

In the presence of numerous possible configurations, this leads to a highly combinatorial optimization problem that is computationally expensive (Onwubolu and Mutingi 2001).

1.2.2 Assembly Line Balancing

In assembly line balancing, individual work elements or tasks are assigned to workstations so that unit assembly cost is minimized as much as possible (Scholl 1999; Sabuncuoglu et al. 2000a, b; Scholl and Becker 2006). Line balancing decisions have a direct impact on the cost-effectiveness of a production process. As such, it is of utmost importance to develop optimal or near-optimal practical solution procedures that can assist decision makers in assembly line balancing decisions, yet with minimal computational requirements.

Figure 1.2 shows a diagraph for a typical line balancing problem and its group representation. The problem consists of 6 different tasks to be allocated to 3 workstations. As illustrated, task groups (1,2), (3,4), and (4,6) are allocated to workstations 1, 2, and 3, respectively. Howbeit, there are several possible solutions that may be generated. The problem becomes highly combinatorial, demanding high computation requirements.

Fig. 1.1 A cellular manufacturing system and its group representation

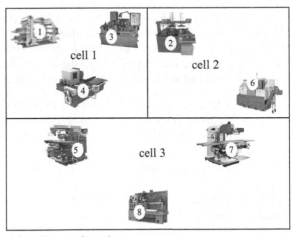

(a) Cell manufacturing system

Machines	1,3,4	2,6	5,7,8
Cells	1	2	3

(b) Group structure

Fig. 1.2 A typical line balancing problem and its grouping representation

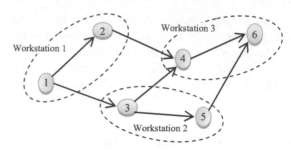

(a) Line balancing problem

Tasks	1,2	3,5	4,6
Workstations	1	2	3

(b) Grouping representation

Due to the combinatorial nature of the assembly line balancing problem, metaheuristic algorithms, such as genetic algorithm, tabu search, simulated annealing, and particle swarm intelligence, are the most viable solution methods to the problem. Iterative metaheuristic approaches can offer reliable solutions within reasonable computational times.

1.2.3 Job Shop Scheduling

The assignment of tasks in a flexible job shop problem environment is more challenging than the classical job shop problem. This requires proper selection of machines from a set of given machines to process each operation. The problem is best defined by three important features: the set of jobs, the set of machines, and the flexibility specification. These are defined as follows:

1. Jobs. $J = \{J_1, J_2, \ldots, J_n\}$ is a set of n independent jobs to be scheduled. Each job J_i consists of a sequence operation to be performed one after the other according to a given sequence. All jobs are assumed to be available at time 0.
2. Machines. $M = \{m_1, m_2, \ldots, m_m\}$ is a set of m machines. Every machine processes only one operation at a time. All machines are assumed to be available at time 0.
3. Flexibility specification: The problem generally falls into two categories, namely: (a) total flexibility, where each operation can be processed on any of the machines; and (b) partial flexibility, where each operation can be performed only on a subset of the machines.

Further to the above definition of the key features, the following simplifying assumptions are essential for formulation of the flexible job shop scheduling environment:

1. Every job operation is performed on one and only one machine at any given point in time;
2. The processing times of the operations are machine-dependent, and the machines are independent from each other;
3. Once started, each operation must be performed to completion without interruption;
4. Setup times are of machines are included in the job processing time of the job; and
5. The transportation time of jobs between machines are negligible and are included in the processing times.

Table 1.2 illustrates a job shop problem with 3 machines to process 3 jobs upon which 8 operations are to be performed. For instance, operation 11, 12, and 13 represent the first, second, and third operations of job 1, respectively. Machines capable of performing each operation are as shown. The objective is twofold: (i) to assign each operation to an appropriate machine, which is a routing problem, and (ii) to sequence the operations on specific machines, which is a sequencing problem, so that make-span and the total working time of machines (total workload) are minimized. This becomes a complex multi-criteria optimization problem that demands significant computational resources.

Figure 1.3 represents a typical solution to the problem, where operations 11, 13, and 31 are performed on machine m_1, operations 21, 23, and 12 are performed on m_2, while 22 and 32 are done on m_3. A group representation scheme of the solution is shown in (b).

Figure 1.3 represents a typical solution to the problem, where operations 11, 13, and 31 are performed on machine m1, operations 21, 23, and 12 are performed on m2, while 22 and 32 are done on m3. A group representation scheme of the solution is shown in (b).

1.2.4 Vehicle Routing Problem

In transportation and distribution, planning for vehicle routing is a major challenge to decision makers in the logistics industry. In order to provide cost-effective and satisfactory delivery (and pickup) services to customers, it is important to optimize the routing of vehicles (Taillard 1999; Mutingi and Mbohwa 2012, b). The vehicle routing problem (VRP) is a hard combinatorial problem that seeks to assign a set of

Table 1.2 A job shop problem with 3 machines and 3 jobs with 8 operations

Jobs	J_1			J_2			J_3	
Operations	11	12	13	21	22	23	31	32
Machines	m_1, m_2, m_3	m_1, m_2	m_1, m_2, m_3	m_1, m_2, m_3	m_1, m_3	m_1, m_2, m_3	m_1, m_2, m_3	m_2, m_3

Fig. 1.3 A typical job shop
problem with 3 machines and
3 jobs

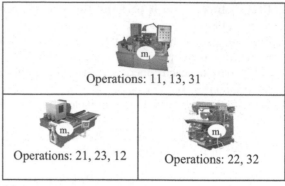

	Operations: 11, 13, 31	
Operations: 21, 23, 12		Operations: 22, 32

(a) Job shop operations

Operations	11,13,31	21,23,12	22,32
Machines	m_1	m_2	m_3

(b) Group structure

groups of customers to a set of vehicles or drivers, so as to minimize the total costs
incurred in visiting all the customers, subject to a number of constraints, such as
follows: (i) use no more vehicles than those available; (ii) satisfy customer demand;
(iii) visit each customer exactly once; (iv) vehicle routes start and finish at the
depot; and (v) vehicle capacity is not violated. In most cases, the main objective is
to minimize the total delivery costs incurred (Gendreau et al. 1999a, b; Mutingi
2013).

Figure 1.4 illustrates a typical vehicle routing schedule, where part (a) shows a
diagraph for the problem, and part (b) shows the group representation of the
problem. The nodes represent 7 customer locations, and the arcs represent the
distances between the customers, and from the depot (denoted by node 0). The
schedule comprises 7 customers that are assigned to 3 available vehicles or drivers.
Further, customer groups (1, 2), (3, 4, 5), and (6,7) are assigned to vehicles v_1, v_2,
and v_3, respectively. The sequence of customers in each group represents the
sequence of customer visits or the direction of the route. This situation is similar to
routing of healthcare workers in a homecare environment (Mutingi and Mbohwa
2013a, b) and handicapped person transportation problems (Rekiek et al. 2006).

1.2.5 Home Healthcare Worker Scheduling

As aging populations continue to increase in most countries, healthcare authorities
continue to face increasing demand for home-based medical care. Most families
prefer their patients to be treated at their homes, rather than at retirement homes and
hospitals (Mutingi and Mbohwa 2013a, b). With the ever-increasing demand for

Fig. 1.4 A typical vehicle
routing schedule

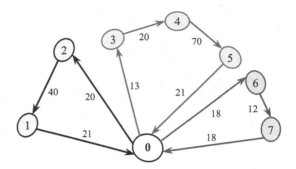

(a) A diagraph for a vehicle routing problem

Operations	1,2	3,4,5	6,7
Machines	v_1	v_2	v_3

(b) Group structure

home healthcare services, many service providers are struggling to find
cost-effective and efficient schedules to meet the expectations of the customers, the
management goals, as well as the desires of the healthcare workers. Home
healthcare service provides continue to expand. Consequently, healthcare worker
scheduling has become a large-scale combinatorial problem, inundated with a
myriad of constraints that have to be taken into account, including patients' preferences, visiting time windows, or travel times depending on the mode of transport.

Figure 1.5 presents a schematic of the home healthcare worker scheduling
problem. In general, the problem is defined thus a set of m available care workers
are given the responsibility to visit n patients for home-based medical care. Each
caregiver k ($k = 1, 2, ..., m$) is supposed to serve a group of patients, where each
patient j ($j = 1,2..., n$) is to be visited within a given time window defined by
earliest start and latest start times, e_j and l_j, respectively. The aim is to minimize the
overall costs of visiting clients (Mutingi and Mbohwa 2013a, b). In this vein, a
penalty cost is incurred whenever a caregiver reaches the client earlier than e_j or
later than l_j. If a_j denotes the caregiver's arrival time at patient j, and p_e and p_l
denote the unit penalty costs incurred when the caregiver arrives too early or too
late, respectively, then $\max[0, e_j - a_j]$ and $\max[0, a_j - l_j]$ have to be minimized, to
maximize patient satisfaction. Furthermore, worker preferences should be taken
into account, if schedule quality is to be maximized (Mutingi and Mbohwa 2013a,
b). However, since planning for home healthcare schedules is especially computationally expensive, effective and efficient algorithms are very important.

Fig. 1.5 Homecare worker schedule and its group structure

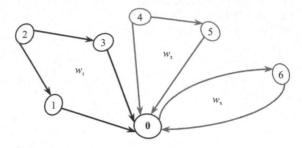

(a) Home healthcare care worker scheduling

patients	1,2,3	4,5	6
workers	w_1	w_2	w_3

(b) Group structure

1.2.6 Bin Packing Problem

In bin packing, objects of different volumes and shapes must be packed into a finite number of bins, where each bin has a specific volume. Oftentimes, the goal is to ensure that the wasted space or number of bins used are minimized as much as possible (Allen et al. 2011; Pillay 2012). According to computational complexity theory, the bin packingproblem is a NP-complete problem that is computationally expensive and highly combinatorial, when formulated as a decision problem (Pillay 2012). Figure 1.6 presents an illustrative example of the bin packing problem and its group structure. In part (a), the problem shows three bins, that is, b_1, b_2, and b_3, which are packed with groups of objects (1, 4, 8), (5, 6), and (2, 3, 7), respectively. This information can be presented conveniently in a group structure as indicated in part (b).

Fig. 1.6 A bin packing problem and its representation

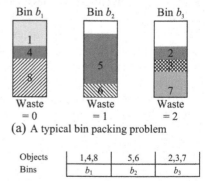

(a) A typical bin packing problem

Objects	1,4,8	5,6	2,3,7
Bins	b_1	b_2	b_3

(b) Group structure

In real-world practice, the bin packing problem comes in different variations, including two-dimensional packing, linear packing, packing by weight, and packing by cost, with many applications. The concepts can be extended to other various situations, such as filling up containers, or loading trucks with weight capacity constraints, metal cutting, and other related problems (Ramesh 2001). In this view, it is important to develop a flexible and adaptive grouping algorithm which can solve this problem and its variants.

1.2.7 Task Assignment Problem

The task assignment problem consists in assigning a set of tasks, $T = \{1, \ldots, n\}$ to an available set of available workers or processors, $W = \{1, \ldots, w\}$ in a manner that will minimize the overall assignment cost function (Tarokh et al. 2011; Mutingi and Mbohwa 2013a, b). The problem is also generally known as a task scheduling problem (Salcedo-Sanz et al. (2006). Basically, each task is defined by its duration and its time window defined by the task's earliest start and latest start times (Mutingi and Mbohwa 2013a, b; Bachouch et al. 2010). Each worker or processor may have a specific scheduled time of day when it is available. In most practical cases, it is desired or required to minimize the workload variation to an acceptable degree, violation of time window constraints, and completion time of all the tasks, depending on the specific situation under consideration.

Figure 1.7 presents an example of a task assignment problem in part (a) and its group structure in part (b). Seven tasks are to be assigned to three processors or workers (called assignees), while minimizing a specific assignment cost function, subject to hard and soft constraints.

The general task assignment problem is a combinatorial problem that is known to be NP-hard due to its myriad of constraints and variables (Salcedo-Sanz et al.

Fig. 1.7 Task assignment and its group representation

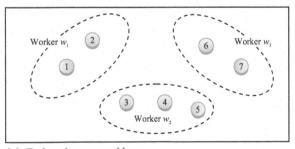

(a) Task assignment problem

Tasks	1,2	3,4,5	6,7
Assignees	w_1	w_2	w_3

(b) Group structure

2006; Chen 2007; Tarokh et al. 2011). It is desirable to develop heuristic algorithms that take advantage of the grouping structure of the problem.

1.2.8 Modular Product Design

Modular design is a design approach that divides a system into modules that can be independently created for assembling a variety of different systems. There are three basic categories of modular design, namely function-based modular design, manufacturing design, and assembly-based modular design (Tseng et al. 2008). Modular design is important as manufacturers need to cope with multiple variations of product specifications and modules in a customized environment (Kreng and Lee 2003). Thus, well-developed modular designsystems will help with the production and control of mass customization. Modular design also focuses on environmental aspects based on grouping techniques.

In the assembly-based modular design method, products are generally described by liaison graph (Tseng et al. 2008). Figure 1.8 shows an example of a Parker Pen assembly with 6 components to be grouped into 3 modules. There are three important stages in modular design:

1. Determination of liaison intensity (LI) of components;
2. Grouping or clustering of components using a grouping method; and
3. Evaluation of the clustering of grouping result.

The goal is to maximize the liaison intensity within each module and to minimize the liaison intensity between modules. However, as the number of components and modules increase, the number of possible combinations increases

Fig. 1.8 A Parker Pen assembly and its group structure

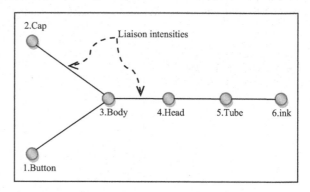

(a) A liaison graph for a parker pen assembly

Components:	1,3,4	5,6	2
Module :	m_1	m_2	m_3

(b) A Group structure for a typical modular design

exponentially, and the complexity of the problem increases rapidly; therefore, developing efficient grouping techniques is imperative (Kamrani and Gonzalez 2003; Tseng et al. 2008).

1.2.9 Group Maintenance Planning

Most industrial systems, such as production systems, pipe networks, mining equipment, aerospace industry, oil and gas, and military equipment, are made up of multi-component systems that require high reliability. Moreover, these systems are normally required to operate with minimal stoppages and breakdowns and, therefore, are usually supported by preventive maintenance/replacement procedures at intervals defined by operating hours in terms of mean time to failure. In the presence of multiple components and subsystems, multiple replacements and maintenance tasks are involved at high costs. Since the reliabilities of the components and subsystems contribute to the overall system reliability, it is essential to formulate the best maintenance strategies for the components and subsystems.

Figure 1.9 shows a typical schedule for system components to be replaced in groups. Part (a) may represent, for example, a set $P = \{P_1, ..., P_{10}\}$ of component pipes in a pipe network that are to be replaced in optimal groups, and the grouping process is supposed to minimize the total costs in terms of distance traveled to repair, preparation, and setup costs. Part (b) is a group structure for the group maintenance schedule, where the 10 pipes are scheduled into 4 groups: $\{1, 5\}$, $\{2, 3, 4, 7\}$, $\{6, 10, 8\}$, and $\{9\}$.

High fixed costs are often incurred in transporting repair equipment to repair facilities and to set them up for the required maintenance procedures. As a result, it is generally more economical to conduct the maintenance or replacements of related

Fig. 1.9 A pipeline maintenance schedule and its group structure

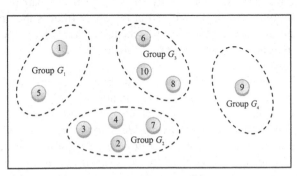

(a) A schedule of pipeline maintenance jobs

Pipes:	1,5	2,3,4,7	6,10,8	9
Groups:	G_1	G_2	G_3	G_4

(b) Group structure for group maintenance schedule

components at one goal. This means that specific groups of components or subsystems have to be cautiously defined so that each group can undergo preventive maintenance within a defined time window. The overall aim is to minimize maintenance costs (setup, preparation costs) while maximizing the reliability of the systems. This grouping problem is twofold (Dekker et al. 1997):

1. Fixed group models, where all components are always jointly maintained as a group; and
2. Optimized-groups models, where several groups are optimally generated, either directly or indirectly.

The major challenge in solving the group maintenance problem is its computational complexity due to exponential growth of the number of variables as the number of components or subsystems increase. Efficient and robust metaheuristic methods are a potential option.

1.2.10 Order Batching

Order batching is a decision problem that is commonplace in warehouse and distribution systems. It is concerned with the search and retrieval of items from their respective storage areas in the warehouse in order to satisfy customer orders. In the real-world practice, customer orders come in small volumes of various types; this makes the retrieval process even more complex. As a result, manual order picking systems need to put in place effective methods to collect items in batches in a more efficient way. Customer orders should be grouped into picking orders of limited sizes, such that the total distance traversed by order pickers is minimized; the total length of all the tours traveled by order pickers should be minimized.

Figure 1.10 shows a plan for three order pickers who are scheduled to pick eight customer orders shown by shaded squares in part (a). The three tours, shown by dotted lines, begin and end at the depot (that is the origin). Part (b) illustrates the group representation for the schedule, where three order pickers O_1, O_2, and O_3 are assigned to pick groups of orders {5, 28, 19}, {45, 51, 85}, and {98, 99}, respectively.

Though the order batching problem can conveniently be modeled based on the group structure, finding the optimal or near-optimal solution poses a computational challenge. Due to the multiplicity of possible combinations, the problem is NP-hard and computationally expensive. However, by taking advantage of the group structure of the problem, a grouping algorithm that iteratively explores improved solutions, while striving to preserve important information in the group structure, may be quite handy.

Fig. 1.10 A tour schedule for order picking and its group representation

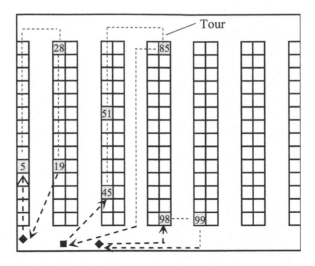

(a) A plan for three order pickers

	5,28,19	45,51,85	98,99
Pickers:	O	O	O

(b) Group structure

1.2.11 Team Formation

Team formation is a recent research area in which a social network of experts is to be organized into teams or groups according to the skills, knowledge, and competences of the experts. The goal is to form the best possible combinations of experts such that the best set of teams is formed to accomplish a given set of projects. The best team formation decision must yield groups of experts in which each team member possesses a greater number of different competences, yet at minimum cost possible. For instance, when auditing a building under construction, auditors with civil and electrical engineering degree, project management skills and other related skills, knowledge, and competencies are certainly required.

Figure 1.11 shows an illustrative example the team formation problem, consisting of a set $S = \{1, 2, ..., 8\}$ of experts and a competence set $C = \{1, 2, ..., 8\}$. Part (a) presents a competence matrix, while part (b) shows its typical solution in form of a diagonalized matrix. Part (c) presents the corresponding group structure. The matrix values are in the range [0, 5] where 0 and 5 denote the lowest and the highest skill levels, respectively, over a particular project or resource (Agustin-Blas et al. 2011a, b). Using iterative permutations of rows and columns, one can obtain the diagonalized matrix which illustrates that the experts can be organized into 3 groups, that is, $\{1, 4\}$, $\{2, 6, 7\}$, and $\{3,5,8\}$.

Fig. 1.11 A team formation competence matrix and its group structure

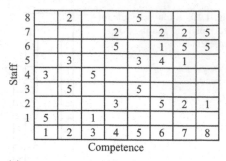

(a) A competence mix matrix

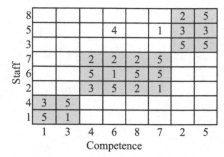

(b) A diagonalized matrix for grouping experts

Experts	1,4	2,6,7	3,5,8
Group	1	2	3

(c) Group structure

The selection of these experts into groups may seem to be easy for small groups. However, we may have a number of experts possessing more than one competence. The complexity of constructing groups of experts in such cases grows exponentially. Approaches in the literature have applied metaheuristics such as fuzzy genetic algorithms in project team formation (Strnad and Guid 2010) and hybrid grouping genetic algorithm for team formation based on group technology (Agustin-Blas et al. 2011a, b).

1.2.12 Earnings Management

Assessing the magnitude of possible earnings management is a difficult and challenging task (Höglund 2013). The most widely used model is linear regression-based model suggested by Jones (1991). Basically, the Jones model separates the non-discretionary (uncontrollable) and the discretionary accruals (controllable). Thus, the discretionary accruals are used as a surrogate for earnings management. The model is built upon the assumption that earnings are managed through accounting accruals. In most applications, companies under study are grouped according to industry types, assuming that the accrual generating process within a specific industry is the same. However, recent studies have shown that this assumption may not hold and may lead to measurement errors (Dopuch et al. 2012). Instead, other grouping variables such as lagged total assets may be used (Ecker et al. 2011). It is necessary to search for alternative methods of grouping companies when applying the Jones model.

It is generally impossible to perform an exhaustive search for the best possible grouping, due to the large number of possible combinations. Even with moderate data sets, the solution space is usually too large for exhaustive search (Höglund 2013). Metaheuristic algorithms are capable of proving optimal or near-optimal solutions within short computation times. Genetic algorithms have been applied successfully in accounting related problems such as in generating bankruptcy prediction rules (Back et al. 1996; Shin and Lee 2002), and in detecting of financial statement fraud (Hoogs et al. 2007).

Figure 1.12 shows an exemplary grouping representation for 40 companies into 5 groups. The task is to group the data set companies and then measure the performance of the cross-sectional Jones model based on the standard deviation of the discretionary accruals and the ability to detect simulated earnings management. Briefly, the model assumes that the reciprocal of total assets TA_t, change in revenues and gross property ΔREV, plant and equipment PPE are regressed on total accruals TACC. According to Höglund (2013), the total accruals are estimated as operating cash flows less net earnings before extraordinary items. All variables in the regression equation are deflated by lagged total assets (TA_{t-1}). Thus, the reciprocal of (TA_{t-1}), ΔREV and PPE are regressed on total accruals (TACC), according to the following expression;

$$\frac{TACC_t}{TA_{t-1}} = \beta_0 \frac{1}{TA_{t-1}} + \beta_1 \frac{\Delta REV_t}{TA_{t-1}} + \beta_2 \frac{PPE_t}{TA_{t-1}} + \varepsilon \quad (1.1)$$

where the residuals from the regressions ε are equivalent to the discretionary accruals. Here, the goodness of a particular grouping solution can be calculated as a function of the standard deviation of the ε values for all the groups. A global grouping algorithm can be more a more effective decision support system for earnings management, even from a multi-criteria perspective.

Companies	12,6,5,7,12,16,22,30	13,9,10,4,8,17,21,31	...	25,2,10,11,14,15,18
Group	1	2	..	5

Fig. 1.12 Group structure for the discretionary accrual estimation model

1.2.13 Economies of Scale

This type of grouping problem occurs in a production environment where a number of products can be produced through a number of processes. A list of customer orders is known a priori, and the orders normally specify a number of possible processes that are capable of produce the orders. In such environments, selecting the right set of processes to execute the orders is a non-trivial task, especially as the number of orders and processes increase. An example of this problem occurs in forging process (Falkenauer 1994), supplier or contractor selection, and related problems.

Figure 1.13 illustrates the economies of scale problem in part (a) and its group representation in part (b). The columns correspond to the available processes, while the rows correspond to the orders. A square with a tick signifies that the process is selected to produce the order, while one without a tick implies that, though feasible, the process is not selected. In this setting, for example, order 1 can be produced through processes 2 and 4; however, process 2 is selected. Likewise, order 8 can be executed by processes 1, 3, and 5; however, process 1 is selected. This can be represented in a group structure as shown in (b). It can be seen that processes 3 and 4 are excluded from this plan.

The aim is to minimize the overall cost incurred in setups and preparations of the processes. These costs correspond to the variable costs of switching from one process to the other, the fixed costs associated with each order/process combination (Falkenauer 1994), and the costs of producing very low quantities (uneconomic). Thus, the orders have to be grouped into large economic batches that should be manufactured by a set of processes, such that the sum of setup and production costs is minimized. An efficient and effective grouping algorithm is crucial for decision support in such environments.

Fig. 1.13 An economies of scale problem and its group structure

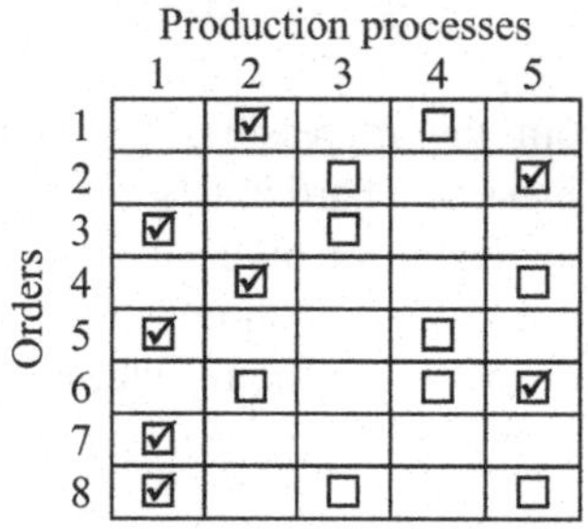

(a) An economies of scale problem

Process:	3,5,7,8	1,4	2,6
Group :	1	2	5

(b) A group structure for the economies of scale problem

1.2.14   Timetabling

Examination timetabling is a scheduling problem that consists in assigning several examinations into limited time slots or periods, subject to hard and soft constraints (Burke and Newall 2004; Pillay and Banzhaf 2010). Hard constraints are to be satisfied at all times, for instance, a student cannot sit for two or more examinations at the same time. On the contrary, soft constraints may be violated, but at a cost, for example, it is desirable to have large-class examinations scheduled early to allow for early assessment of the examinations and to provide study time for each student between any two examinations (Rakesh et al. 2014). Though soft constraints may be violated at a cost, it is highly desirable to satisfy them as much as possible. It follows that a high-quality timetable should not violate any hard constraints, while satisfying soft constraints as much as possible. Therefore, the objective is to maximize the quality of the timetable as much as possible, while satisfying all soft constraints.

Figure 1.14 shows a snapshot of an examination schedule spanning over 2 days, that is, Day 1 and Day 2. Each day has 3 time slots, 9–12 noon, 1–3 pm, and 4–6 pm. A set of examinations are placed in each time slot, for example, examinations E12 and E9 are in the first time slot (9–12 noon), examination E5 is in the second slot (1–3 pm), and examinations E8, E6, and E2 in the third slot (4–6 pm). In this case, the schedule can be viewed as a grouping problem defined by 6 groups {12, 9}, {5}, {8, 6, 2}, {3, 4}, {7, 11, 10}, and {1, 12}, where the time slots or periods are the group identities containing examinations.

For small-scale problems, the timetabling problem seems tractable. However, as the number of days or time slots, number of examinations, number of possible examination venues, and number of constraints increase, the complexity of the problem grows exponentially. Efficient and adaptable solution approaches are essential for such complex grouping problems.

Day 1			Day 2		
9 - 12	1 - 3	4 - 6	9 - 12	1 - 3	4 - 6
E12	E5	E8	E3	E7	E1
E9		E6	E4	E11	E12
		E2		E10	

(a) An economies of scale problem

Exams:	12,9	5	8,6,2	3,4	7,11,10	1,12
Time slots:	1	2	3	4	5	6

(b) A group structure for the timetabling problem

Fig. 1.14 The timetabling problem and its group structure

1.2.15　Student Grouping for Cooperative Learning

Cooperative learning is an active pedagogy that seeks to enhance the learning performance of students by maximizing the relational or social network of the students (Tsay and Brady 2010; Kose et al. 2010). In real-world practice, students can list a number of preferred classmates with whom they would like to cooperate. An index system can be used, where a "1" stands for the first choice, "2" for the second, and so on, up to a predefined maximum value. Usually, a questionnaire is designed to collect the preferences of the students, which can then be aggregated using a sociometric tool to determine the social indices (Chen et al. 2012a, b). In brief, cooperative learners grouping requires that learners ($i = 1, 2, \ldots, m$), are to be clustered into g groups ($j = 1, 2, \ldots, g$), where learner i can be allocated to any group j, subject to group size limits. This can be represented by a group encoding scheme.

Figure 1.15 demonstrates the cooperative learners grouping problem consisting of 10 learners that are to be partitioned into g groups. Each group should have at least 2 members, and at most 3. Part (b) indicates the group chromosomal representation of a typical solution to this example, where four groups, $j = 1, 2, 3,$ and 4, correspond to groups of learners {3, 7}, {1, 4, 5}, {2, 6, 10}, and {8, 9}, respectively.

The goal of the cooperative learners grouping problem is to maximize the overall social status index for all the whole class. The problem may be tractable over small numbers of students and groups. However, as the numbers increase, the task becomes a complex grouping problem, especially when some side constraints are imposed. Practical methods that can find feasible solutions in shorter computation

Fig. 1.15 The cooperative learners group problem and its group structure

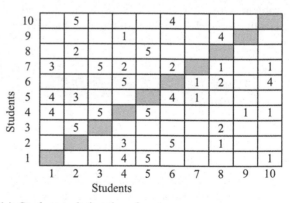

(a) Students relational preferences

Learners:	3,7	1,4,5	2,6,10	8,9
Group:	1	2	3	4

(b) Group structure for cooperative learners

times could be more reliable. One such approach is grouping genetic algorithm, or its variants, which has been successfully applied to complex grouping problems.

1.2.16 Other Problems

Apart from the problems outlined in the previous section, there are other related grouping problems that exist in various industry disciplines, such as districting problem in healthcare (Benzarti 2012), facility location (Pitaksringkarn and Taylor 2005a, b), advertisement allocation (Dao et al. 2012), and data clustering (Nezamabadi-pour and Dowlatshahi 2014), and material cutting plan problem (Hung et al. 2003). Most, if not all, of these problems can be modeled and solved using grouping genetic algorithms.

The next section presents an outline of past approaches to grouping problems found in the literature.

1.3 Extant Modeling Approaches to Grouping Problems

Many NP-hard combinatorial optimization grouping problems have been outlined in this chapter. Most of these problems are highly constrained such that not all possible groupings are permitted; group formations must satisfy a number of constraints. However, it was also noted that the groups are supposed to be formed in accordance with the objective function which is normally derived based on the structure of the groups. It is logical, therefore, that the search mechanisms be centered on the groups themselves (or group segments) rather than the members of the group.

Extant modeling approaches largely depend on the subclass of the grouping problem under study. Grouping problems can be classified into constant grouping problem and variable grouping problems. While constant grouping involves fixed and predetermined size grouping, the variable grouping problem assumes that the group size is not known a priori and the algorithm should determine the most appropriate group size(s). Another most important classification is order-dependent grouping problems and order-independent grouping problems. In order-dependent problems, such as in vehicle routing problem and timetabling problem, the quality of the solutions is influenced by the way in which the group members are sequenced. On the other hand, the quality of solutions in order-independent grouping problems are not affected by the sequence of members in the groups, for example, review group formation, students grouping for cooperative learning, and the general team formation problem. These views affect the choice of modeling approaches to grouping problems.

Past modeling approaches can be classified broadly into statistical clustering, mathematical programming methods, basic problem-specific heuristic methods, and general-purpose metaheuristic methods such as genetic algorithms, particle swarm optimization, ant colony optimization, simulated annealing, and tabu search. Due the fact that grouping problems are NP-hard, metaheuristic approaches have been found to be most appropriate. Falkenauer (1996) realized that classical genetic algorithms (and therefore, related algorithms) generally do not perform well on grouping problems. The main observations from the past applications are as follows:

1. High redundancy in the basic straightforward encoding of the population-based approaches, since the cost function depends on the grouping rather than the individual members;
2. Context insensitivity of algorithm operators, such as crossover function of the genetic algorithm, which may lead to casting context-dependent information out of context;
3. Schema disruption, especially when solving practical grouping problems with long schemata, where algorithm operators (e.g., crossover) may destroy good schema over the search journey.

Unlike the classical approaches, the focus of grouping genetic algorithms is on groups of items. Introduced by Falkenauer (1994), the grouping genetic algorithm (GGA) is the most established metaheuristic for solving grouping problems by exploiting the structural information and the grouping structure of these problems to enhance the search process. It is against this background that this book is focused on the extensions, advances, and applications of the grouping genetic algorithm and its variants. The next section summarizes the overall structure of the book.

1.4   Structure of the Book

This book consists of 4 parts: Part 1 is the introductory section that consists of two chapters. This chapter explores and presents a wide range of grouping problems across various industry disciplines. Chapter 2 presents complicating features behind the process of modeling and solving of the identified grouping problems.

Part 2 describes the grouping genetic algorithm and its variants. Chapter 3 presents the main grouping genetic algorithm, explaining its unique grouping genetic operators that enhance the optimization algorithm. Chapter 4 explains the fuzzy grouping genetic algorithm, a variant of the former that deals with grouping problems in a fuzzy environment.

Part 3 focuses on the research applications of the grouping genetic algorithms and its variants. Presented in this section are various types of grouping problems from different industrial disciplines and the applications of the grouping algorithms. Illustrative computational experiments and discussions are presented to demonstrate the advances and applications of the grouping genetic algorithms.

Part 4 presents concluding remarks and potential areas for possible further research on grouping genetic algorithms and its applications. The next chapter deals with the identification of various complicating features found in grouping problems, which further points to the need for advances in grouping algorithms and their applications.

References

Agustın-Blas LE, Salcedo-Sanz S, Ortiz-Garcıa EG, Portilla-Figueras A, Perez-Bellido AM, Jimenez-Fernandez S (2011a) Team formation based on group technology: a hybrid grouping genetic algorithm approach. Comput Oper Res 38:484–495

Agustın-Blas LE, Salcedo-Sanz S, Vidales P, Urueta G, Portilla-Figueras JA (2011b) Near optimal citywide WiFi network deployment using a hybrid grouping genetic algorithm. Expert Syst Appl 38(8):9543–9556

Allen SD, Burke EK, Kendall G (2011) A hybrid placement strategy for the three-dimensional strip packing problem. Eur J Oper Res 209(3):219–227

Althaus E, Baumann T, Schömer E, Werth K (2007) Trunk pack-ing revisited. LNCS 4525: 420–430

Bachouch RB, Liesp AG, Insa L, Hajri-Gabouj S (2010) An optimization model for task assignment in home healthcare. In: IEEE workshop on health care management (WHCM), pp 1–6

Back B, Laitinen T, Sere K (1996) Neural networks and genetic algorithms for bankruptcy predictions. Expert Syst Appl 11(4):407–413

Badeau P, Guertin F, Gendreau M, Potvin J-Y, Taillard E.D (1997) A parallel tabu search heuristic for the vehicle routing problem with time windows. Transport Res 5C: 109–122

Baker BM, Benn C (2001) Assigning pupils to tutor groups in a comprehensive school. J Oper Res Soc 52:623–629

Bartov E, Gul FA, Tsui JSL (2000) Discretionary-accruals models and audit qualifications. J Account Econ 30(3):421–452

Bektas T (2006) The multiple traveling salesman problem: an overview of formulations and solution procedures. Omega 34(3):209–219

Benzarti E (2012) Home health care operations management: applying the districting approach to home health care. École Centrale Paris, Thesis

Brandao J (2008) A deterministic tabu search algorithm for the fleet size and mix vehicle routing problem. Eur J Oper Res 195(3):716–728

Burke EK, Newall JP (2004) Solving examination timetabling problems through adaptations of heuristic orderings: models and algorithms for planning and scheduling problems. Ann Oper Res 129(1–4):107–134

Carter AE, Ragsdale CT (2006) A new approach to solving the multiple traveling salesperson problem using genetic algorithms. Eur J Oper Res 175(1):246–257

Chen Y-Y (2013) Fuzzy flexible delivery and pickup problem with time windows. Information technology and quantitative management (ITQM2013). Proc Comput Sci 17:379–386

Chen AL, Martinez DH (2012) A heuristic method based on genetic algorithm for the baseline-product design. Expert Syst Appl 39(5):5829–5837

Chen Y, Fan Z-P, Ma J, Zeng S (2011) A hybrid grouping genetic algorithm for the reviewer group construction problem. Expert Syst Appl 38:2401–2411

Chen JC, Wu C-C, Chen C-W, Chen K-H (2012a) Flexible job shop scheduling with parallel machines using genetic algorithm and grouping genetic algorithm. Expert Syst Appl 39 (2012):10016–10021

Chen R-C, Chen S-Y, Fan J-Y, Chen Y-T (2012b) Grouping partners for cooperative learning using genetic algorithm and social network analysis. The 2012 International Workshop on Information and Electronics Engineering (IWIEE). Proc Eng 29:3888–3893

Cheng M, Ozaku HI, Kuwahara N, Kogure K, Ota J (2007) Nursing care scheduling problem: analysis of staffing levels. In: IEEE proceedings of the 2007 international conference on robotics and biomimetics, vol 1, pp 1715–1719

Dao TH, Jeong SR, Ahn H (2012) A novel recommendation model of location-based advertising: Context-aware collaborative filtering using GA approach. Expert Syst Appl 39(2012): 3731–3739

de Jonge B, Klingenberg W, Teunter R, Tinga T (2016) Reducing costs by clustering maintenance activities for multiple critical units. Reliab Eng Syst Saf 145:93–103

De Lit P, Falkenauer E, Delchambre A (2000) Grouping genetic algorithms: an efficient method to solve the cell formation problem. Math Comput Simul 51(3–4):257–271

Dekker R, Wildeman RE, Van Der Duyn SFA (1997) A review of multi-component maintenance models with economic dependence. Math Methods Oper Res 45:411–435

Dereli T, Baykasoglu A, Das GS (2007) Fuzzy quality-team formation for value added auditing: a case study. J Eng Tech Manage 24(4):366–394

Dopuch N, Mashruwala R, Seethamraju C, Zach T (2012) The impact of a heterogeneous accrual-generating process on empirical accrual models. J Account, Auditing and Financ 27(3): 386–411

Ecker F, Francis J, Olsson P, Schipper K (2011) Peer firm selection for discretionary accruals models. Duke University Working Paper

Falkenaeur E (1993) The grouping genetic algorithms—widening the scope of the GAs. JORBEL Belg J Oper Res Stat Comput Sci 33:79–102

Falkenaeur E (1995) Solving equal piles with a grouping genetic algorithm. In: LJ. Eschelman (ed) Proceedings of the sixth international conference on genetic algorithms (IGGA95). San Mateo, CA, July 1995. University of Pittsburg (Pennsylvania), Morgan Kaufman Publishers, pp 492–497

Falkenauer E (1992) The grouping genetic algorithms—widening the scope of the GAs. Belg J Oper Res Stat Comput Sci 33:79–102

Falkenauer E (1994) A new representation and operators for GAs applied to applied to grouping problems. Evol Comput 2:123–144

Falkenauer E (1996) A hybrid grouping genetic algorithm for bin packing. J Heuristics 2:5–30

Filho EVG, Tiberti AJ (2006) A group genetic algorithm for the machine cell formation problem. Int J Prod Econ 102:1–21

Gendreau M, Laporte G, Musaragany C, Taillard ED (1999a) A tabu search heuristic for the heterogeneous fleet vehicle routing problem. Comput Oper Res 26(12):1153–1173

Gendreau M, Laporte G, Musaraganyi C, Taillard ED (1999b) A tabu search heuristic for the heterogeneous fleet vehicle routing problem. Comput Oper Res 26:1153–1173

Gunn EA, Diallo C (2015) Optimal opportunistic indirect grouping of preventive replacements in multicomponent systems. Comput Ind Eng 90:281–291

Henn S (2012) Algorithms for on-line order batching in an order picking warehouse. Comput Oper Res 39:2549–2563

Henn S, Wäscher G (2012) Tabu search heuristics for the order batching problem in manual order picking systems. Eur J Oper Res 222:484–494

Hettich S, Pazzani MJ (2006) Mining for element reviewers: lessons learned at the national science foundation. In Proceeding of the KDD'06, Philadelphia, Pennsylvania, USA, pp 862–871

Ho GTS, Ip WH, Lee CKM, Mou WL (2012) Customer grouping for better resources allocation using GA based clustering technique. Expert Syst Appl 39:1979–1987

Hoogs B, Kiehl T, Lacomb C, Senturk D (2007). A genetic algorithm approach to detecting temporal patterns indicative of financial statement fraud. Intell Syst Account, Financ Manage 15(1–2): 41–56

Höglund H (2013) Estimating discretionary accruals using a grouping genetic algorithm. Expert Syst Appl 40:2366–2372

Hung CY, Sumichrast RT, Brown EC (2003) CPGEA: a grouping genetic algorithm for material cutting plan generation. Comput Ind Eng 44:651–672

Jones JJ (1991) Earnings management during import relief investigation. J Account Res 29(2): 193–228

Joung Y-K, Noh SD (2014) Intelligent 3D packing using a grouping algorithm for automotive container engineering. Journal of Computational Design and Engineering 1(2):140–151

Kaaouache MA, Bouamama S (2015) Solving bin packing problem with a hybrid genetic algorithm for VM Placement in Cloud. Proc Comput Sci 60:1061–1069

Kamrani AK, Gonzalez R (2003) A genetic algorithm-based solution methodology for modular design. J Intell Manuf 4(6):599–616

Kivelevitch E, Cohen K, Kumar M (2013) A market-based solution to the multiple traveling salesmen problem. J Intell Rob Syst 72(1):21–40

Kose S, Sahin A, Ergu A, Gezer K (2010) The effects of cooperative learning experience on eight grade students' achievement and attitude toward science. Education 131(1):169–180

Kreng VB, Lee T-P (2003) Product family design with grouping genetic algorithm—a case study. J Chin Inst Ind Eng 20(4):373–388

Kreng VB, Lee T-P (2004) Modular product design with grouping genetic algorithm—a case study. Comput Ind Eng 46(3):443–460

Landa-Torres I, Gil-Lopez S, Del Ser J, Salcedo-Sanz S, Manjarres D, Portilla-Figueras JA (2013) Efficient citywide planning of open WiFi access networks using novel grouping harmony search heuristics. Eng Appl Artif Intell 26(2013):1124–1130

Li F, Sun Y; Ma L, Mathew J (2011) A grouping model for distributed pipeline assets maintenance decision. In: The 2011 international conference on quality, reliability, risk, maintenance, and safety engineering (ICQR2MSE), 17–19 June 2011, pp 601–606

Liu S, Huang W, Ma H (2009a) An effective genetic algorithm for the fleet size and mix vehicle routing problems. Transp Res Part E 45:434–445

Luh PB, Gou L, Zhang Y, Nagahara T, Tsuji M (1998) Job shop scheduling with group-dependent setups, finite buffers, and long time horizon. Ann Oper Res 76:233–259

Moghadam B, Seyedhosseini S (2010) A particle swarm approach to solve vehicle routing problem with uncertain demand: A drug distribution case study. Int J Ind Eng Computations 1(1): 55–64

Mutingi M (2013) Group Genetic Algorithms for Heterogeneous Vehicle Routing. In Masegosa AD, Villacorta PJ, Cruz-Corona C, García-Cascales MS, Lamata MT, Verdegay JS (ed) Exploring innovative and successful applications of soft computing. IGI-Global, USA, pp 161–180

Mutingi M, Mbohwa C (2012a) Enhanced group genetic algorithm for the heterogeneous fixed fleet vehicle routing problem. In: IEEE conference on industrial engineering and engineering management. Hong Kong, Dec

Mutingi M, Mbohwa C (2012b) Group genetic algorithm for the fleet size and mix vehicle routing problem. In: Proceedings of the IEEE international conference on industrial engineering and engineering management. Hong Kong, 10–13 Dec 2012

Mutingi M, Mbohwa C (2013a) Task assignment in home health care: a fuzzy group genetic algorithm approach. In: The 25th annual conference of the Southern African institute of industrial engineering 2013. Stellenbosch, South Africa, 2013, 9–11 July, p 6341

Mutingi M, Mbohwa C (2013b) Home healthcare worker scheduling: a group genetic algorithm approach. In: Proceedings of the world congress on engineering 2013. London, UK, 3–5 July 2013, pp 721–725

Mutingi M, Mbohwa C (2014a) A fuzzy-based particle swarm optimization approach for task assignment in home healthcare. S Afr J Ind Eng 25(3):84–95

Mutingi M, Mbohwa C (2014b) Multi-objective homecare worker scheduling—a fuzzy simulated evolution algorithm approach. IIE Trans Healthc Syst Eng 4(4):209–216

Mutingi M, Mbohwa C (2014c) A fuzzy grouping genetic algorithm for care task assignment. In: Proceedings of the world congress on engineering and computer science. San Francisco, USA, 22–24 Oct 2014

Mutingi M, Onwubolu GC (2012) Integrated cellular manufacturing system design and layout using group genetic algorithms. In: Aziz FA (ed) Manufacturing system. InTech-Open, USA, pp 205–222

Mutingi M, Mbohwa C, Mhlanga S, Goriwondo W (2012) Integrated cellular manufacturing system design: an evolutionary algorithm approach. In: Proceedings of the 3rd international conference on industrial engineering and operations management. Turkey, 3–6 July 2012, pp 254–264

Nezamabadi-pour MB, Dowlatshahi H (2014) GGSA: a grouping gravitational search algorithm for data clustering. Eng Appl Artif Intell 36:114–121

Onwubolu GC, Mutingi M (2001) A genetic algorithm approach to cellular manufacturing systems. Comput Ind Eng 39:125–144

Onwubolu GC, Mutingi M (2003) A genetic algorithm approach for the cutting stock problem. J Intell Manuf 14:209–218

Parragh SN, Doerner HF, Hartl RF (2008) A survey on pickup and delivery problems part II: transportation between pickup and delivery locations. J Betriebswirtschaft 58(2008):81–117

Phanden RK, Jain A, Verma R (2012) A genetic algorithm-based approach for job shop scheduling. J Manuf Technol Manage 23(7):937–946

Pillay N (2012) A study of evolutionary algorithm selection hyper-heuristics for the one-dimensional binpacking problem. S Afr Comput J 48:31–40

Pillay N, Banzhaf W (2010) An informed genetic algorithm for the examination timetabling problem. Appl Soft Comput 10:457–467

Pitaksringkarn L, Taylor MAP (2005a) Grouping genetic algorithm in GIS: a facility location modelling. J East Asia Soc Transp Stud 6:2908–2920

Pitaksringkarn L, Taylor MAP (2005b) Grouping genetic algorithm in GIS: a facility location modelling. J East Asia Soc Transp Stud 6:2908–2920

Potvin J-Y, Bengio S (1996) The vehicle routing problem with time windows part II: genetic search. J Comput 8(2): 165–172

Rakesh P, Badoni DK, Gupta Pallavi Mishra (2014) A new hybrid algorithm for university course timetabling problem using events based on groupings of students. Comput Ind Eng 78:12–25

Ramesh R (2001) A generic approach for nesting of 2-D parts in 2-D sheets using genetic and heuristic algorithms. Comput Aided Des 33(12):879–891

Rekiek B, De Lit P, Pellichero F, Falkenauer E, Delchambre A (1999) Applying the equal piles problem to balance assembly lines. In: Proceedings of the 1999 IEEE international symposium on assembly and task planning Porto. Portugal, July 1999

Rekiek B, Delchambre A, Saleh HA (2006) Handicapped person transportation: an application of the grouping genetic algorithm. Eng Appl Artif Intell 19(5):511–520

Renaud J, Boctor FF (2002) A sweep-based algorithm for the fleet size and mix vehicle routing problem. Eur J Oper Res 140:618–628

Rochat Y, Taillard ED (1995) Probabilistic diversification and intensification in local search for vehicle routing. J Heuristics 1: 147–167

Rostom MR, Nassef AO, Metwalli SM (2014) 3D overlapped grouping GA for optimum 2D guillotine cutting stock problem. Alexandria Eng J 53(3):491–503

Rubinovitz J, Levitin G (1995) Genetic algorithm for assembly line balancing. Int J Prod Econ 41(1–3):343–354

Sabuncuoglu I, Erel E, Tanyer M (2000a) Assembly line balancing using genetic algorithms. J Intell Manuf 11(3):295–310

Sabuncuoglu I, Erel E, Tanyer M (2000b) Assembly line balancing using genetic algorithms. J Intell Manuf 11(3):295–310

Salcedo-Sanz S, Xu Y, Yao X (2006) Hybrid meta-heuristics algorithms for task assignment in heterogeneous computing systems. Comput Oper Res 33:820–835

Scholl A (1999) Balancing and sequencing of assembly lines. Physica-Verlag, Heidelberg

Scholl A, Becker C (2006) State-of-the-art exact and heuristic solution procedures for simple assembly line balancing. Eur J Oper Res 168:666–693

Sheu JB (2007) A hybrid fuzzy-optimization approach to customer grouping based logistics distribution operations. Appl Math Model 31(6): 1048–1066

Shin KS, Lee YJ (2002). A genetic algorithm application in bankruptcy prediction modeling. Expert Syst Appl 23(3): 321–328

Strnad D, Guid N (2010) A Fuzzy-Genetic decision support system for project team formation. Appl Soft Comput :1178–1187

Taillard ED (1999) A heuristic column generation method for the heterogeneous fleet VRP. RAIRO 33:1–34

Tarantilis CD, Kiranoudis CT, Vassiliadis VSA (2003) A list based threshold accepting metaheuristic for the heterogeneous fixes fleet vehicle routing problem. J Oper Res Soc 54(1):65–71

Tarantilis CD, Kiranoudis CT, Vassiliadis VSA (2004) A threshold accepting metaheuristic for the heterogeneous fixed fleet vehicle routing problem. Eur J Oper Res 152:148–158

Tarokh MJ, Yazdanib M, Sharifi M, Mokhtari MN (2011) Hybrid meta-heuristic algorithm for task assignment problem. J Optim Ind Eng 7:45–55

Tsay M, Brady M (2010) A case study of cooperative learning and communication pedagogy: Does working in teams make a difference? J Sch Teach Learn 10(2):78–89

Tseng H-E, Chang CC, Li J-D (2008) Modular design to support green life-cycle engineering. Expert Syst Appl 34:2524–2537

Tutuncu GY (2010) An interactive GRAMPS algorithm for the heterogeneous fixed fleet vehicle routing problem with and without backhauls. Eur J Oper Res 201(2):593–600

Van Do P, Barros A, Bérenguer C, Bouvard K, Brissaud F (2013) Dynamic grouping maintenance with time limited opportunities. Reliab Eng Syst Saf 120:51–59

Wang HF, Chen Y-Y (2013) A coevolutionary algorithm for the flexible delivery and pickup problem with time windows. Int J Prod Econ 141(1):4–13

Wessner M, Pfister H-R (2001) Group formation in computer-supported collaborative learning. In: Proceedings of the 2001 international ACM SIGGROUP conference on supporting group work, pp 24–31

Wi H, Oh S, Mun J, Jung M (2009) A team formation model based on knowledge and collaboration. Expert Syst Appl 36(5):9121–9134

Yanık S, Sürer Ö, Öztayşi B (2016) Designing sustainable energy regions using genetic algorithms and location-allocation approach. Energy 97:161–172

Yu S, Yang Q, Tao J, Tian X, Yin F (2011) Product modular design incorporating life cycle issues —group genetic algorithm (GGA) based method. J Clean Prod 19(9–10):1016–1032

Chapter 2
Complicating Features in Industrial Grouping Problems

2.1 Introduction

Real-world grouping problems are generally inundated with several complicating features that pose serious challenges to decision makers in various sectors of industry. Some of these real-world problems are estimation of discretion accruals, customer grouping, modular product design, order batching, group maintenance planning, flexible job scheduling, team formation, homecare worker scheduling, and timetabling, to mention but a few. A lot more of these problems and their variants are found in industry.

Recently, the growth of the literature on the subject generally has indicated that more large-scale real-world cases of the problem continue to appear worldwide. Furthermore, researchers in this area continue to discover more and more complexities associated with the problems. On the other hand, the general growth of global competition in the global marketplace has left the decision makers with no option except to rethink and reinvent their ways of business. For instance, a number of grouping problems in logistics are associated with the urgent need for efficient and cost-effective methods of optimizing logistics operations. Without that, the quality of service and customer delivery will be affected severely, leading to loss of business in the medium to long term. Customers and other stakeholders have become aware of the best services they deserve, and as a result, their expectations continue to grow. In the same vein, the customer (and stakeholder) voice is not only calling for the fulfillment of her basic needs, but also for products and processes that take care of the society, the environment, the economy, and energy. Such issues have led to research in home healthcare scheduling, green modular product design, green vehicle routing, customer grouping, economies of scale, and estimation of discretionary accruals, among others. The same is common across several other disciplines in industry. Therefore, developing efficient and effective tools for decision support in grouping problems is quite imperative.

© Springer International Publishing Switzerland 2017
M. Mutingi and C. Mbohwa, *Grouping Genetic Algorithms*,
Studies in Computational Intelligence 666,
DOI 10.1007/978-3-319-44394-2_2

Learning from literature search surveys, grouping problems are highly combinatorial and NP-hard (Falkenauer 1998); they cannot be solved to optimality in polynomial time. It will be interesting to learn from real-world case studies in the literature, so as to unearth the various challenges and complicating features behind their complexity. One would ask these questions: What are the common complicating features behind grouping problems? How do these features contribute to the complexity of the problems? What are the most robust and reliable approaches for solving the problems?

Due to the inherent challenges experienced in modeling and solving grouping problems in industry, several researchers have attempted to suggest approaches that can address the complexities of the problems in a more efficient and effective manner. This has stimulated the search for flexible and efficient approximate approaches that can handle the problems while providing near-optimal or optimal solutions. These approaches include not only heuristics for particular classes of problems, but also the use of general-purpose metaheuristics which are capable of solving hard problems by exploring large solution spaces through effective reduction of the size of solution space and efficient exploitation of the reduced space. Some of these methods include simulated annealing (Matusiak et al. 2014), particle swam optimization (Moghadam and Seyedhosseini 2010; Mutingi and Mbohwa 2014), genetic algorithms (Onwubolu and Mutingi 2003; Mutingi and Mbohwa 2013), tabu search (Gendreau et al. 1999), ant colony optimization (Nothegger et al. 2012), simulated metamorphosis (Mutingi and Mbohwa 2015), and other evolutionary algorithms (Ochi et al. 1998; Prins et al. 2004; Hindi et al. 2002; Cote et al. 2004; Santiago-Mozos et al. 2005).

Notable remarkable work in this area was done by Falkenauer (1996), in which a grouping genetic algorithm was developed especially for solving grouping problems more efficiently. The algorithm was motivated by the rising challenges and complexities in grouping problems and the realization of the inadequacies of the conventional genetic algorithms (and also other related algorithms). In Falkenauer's view, conventional genetic algorithms have three major limitations in regard to grouping problems:

1. The traditional type of chromosome encoding wastes a significant redundant space;
2. It is not easy to generate good quality offspring through the standard reproduction mechanism such as Roulette wheel; and
3. The conventional genetic operators such as crossover and mutation tend to spoil the quality of the offspring population.

As noted earlier, since the inception of the grouping genetic algorithm, the challenges, complexity, and variety of grouping problems have continued to grow. This necessitates further research on the advances and applications of grouping genetic algorithm and its variants. As such, the purpose of this chapter is to obtain further understanding of the common complicating features inherent in industrial grouping problems. By the end of the chapter, the following learning outcomes are expected:

1. To know the various existing real-world cases of industrial grouping problems in the literature;
2. To understand the taxonomic classification of the complicating features behind grouping problems; and
3. To evaluate and recommend suitable solution approaches to real-world grouping problems.

The remainder of the chapter is organized as follows: The next section outlines the research methodology, which essentially focuses on literature search survey of existing case studies in the literature. Section 2.3 summarizes the research findings. A taxonomic discussion on the complicating features is presented in Sect. 2.4. Suggested solution approaches are outlined in Sect. 2.5. Finally, Sect. 2.6 summarizes the chapter.

2.2   Research Methodology

To obtain an in-depth understanding of the nature and characteristics of grouping problems, recent case studies were selected from the literature. A wide range of online literature sources and databases were used to compile the case studies. To a large extent, databases such as ScienceDirect, EBSCO Inspec, ISI Web of Science, and Ei Compendex were utilized. The key words that were used in the search process include "grouping problem," "clustering problem," "grouping algorithm" and "group technology," and "group allocation." The purpose of using key words was to streamline irrelevant studies and to eliminate some of the studies whose major focus was not related to grouping problems. Figure 2.1 provides a summary of the research methodology for the study.

Major case studies were selected, and the complicating features were identified from the studies. Solution approaches that were applied to the case studies were also noted. The next section presents the research findings.

2.3   Research Findings

The focus of the search survey was on recent literature, beginning from the year 2000. This enabled the research to obtain the most appropriate and current state-of-the-art information and draw relevant conclusions from recent cases in the literature. Furthermore, efforts were made to gather case studies from various disciplines in industry, to further substantiate the validity of the conclusions drawn. A summary of the results of the literature search survey is presented in Table 2.1.

Fig. 2.1 Research
methodology

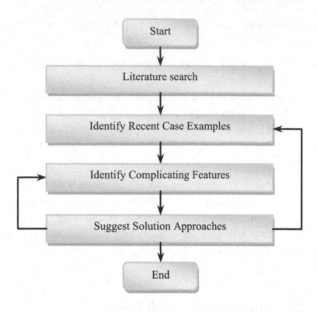

Table 2.1 Selected recent case studies

No.	Grouping problem	Solution approach	References
1	Modular product design of a family of three general aviation aircraft in USA	Genetic algorithm	D'Souza and Simpson (2003)
2	Modular product design for an electrical consumer product provided by an original design manufacturer in Taiwan	Grouping genetic algorithm	Kreng and Lee (2004)
3	Customer grouping for resource allocation at a window curtain manufacturer based on China	Genetic algorithm-based k-means clustering technique	Ho et al. (2012)
4	Estimating discretionary accruals in the USA	Grouping genetic algorithm	Hoglund (2013)
5	Facility location modeling for agricultural logistics in Thailand	Grouping genetic algorithm in GIS	Pitaksringkarn and Taylor (2005)
6	Group maintenance for a company contracted to maintain engines in the aerospace industry in Canada	Network tree formulation; depth-first shortest path algorithm	Gunn and Diallo (2015)
7	Flexible job scheduling for a weapon production factory in Taiwan	Genetic algorithm, grouping genetic algorithm	Chen et al. (2012)
9	Heterogeneous student grouping for Bangkok University	Genetic algorithm for heterogeneous grouping	Sukstrienwong (2012)

(continued)

Table 2.1 (continued)

No.	Grouping problem	Solution approach	References
10	Intelligent 3D container loading for automotive container engineering in South Korea	Intelligent packing algorithm	Joung and Noh (2014)
11	Team formation based on group technology with real application in a Spanish University, Spain	Group technology, hybrid grouping genetic algorithm	Agustín-Blas et al. (2011)
12	Reviewer group construction for National Science Foundation of China (NSFC), China	Hybrid grouping genetic algorithm	Chen et al. (2011)
13	Order batching for precedence constrained orders for a large order picking warehouse in Finland	Simulated annealing	Matusiak et al. (2014)
14	Timetabling for 13 real-world problems —Carter Benchmarks	Informed genetic algorithm	Pillay and Banzhaf (2010)
15	Maintenance grouping model for an industrial case study in Australia	A modified genetic algorithm	Li et al. (2011a)
16	Homecare worker scheduling for community care service in the UK	Particle swarm optimization algorithm	Akjiratikarl et al. (2007)
17	Timetabling for a University of Agriculture in Nigeria	Genetic algorithm	Arogundade et al. (2010)
18	Vehicle (homogenous) routing with prioritized time windows for a distribution company in Iran	Cooperative coevolutionary multi-objective quantum-genetic algorithm	Beheshti et al. (2015)
19	Subcontractor selection for construction industry in Turkey	Genetic algorithms	Polat et al. (2015)

Following the analysis of the literature search survey, a total of 18 recent case studies from different disciplines were selected. Interesting observations were realized from the case studies. It was observed that the cases were conducted in different places, including USA, UK, Canada, Australia, Spain, Thailand, Taiwan, China, Iran, South Korea, and Nigeria. This indicates the widespread of research activities in grouping problems across the world. Furthermore, the range or variety of the problem types is an indication of the increasing realization of grouping problems in various disciplines.

It is also interesting to note that all the cases were solved using metaheuristic approaches. This can be attributed to the growing complexity and hardness of the grouping problem. Furthermore, it can be seen that about 72.2 % utilize genetic algorithm or a variant of the same. This tends to indicate that there is more potential in the use of genetic algorithm-based approaches.

A closer look at the case studies revealed a number of common complicating features among grouping problems. Figure 2.2 shows a taxonomic analysis of the complicating factors in industrial grouping problems, classifying the problems into

Fig. 2.2 A taxonomic
analysis of complicating
features in industrial grouping
problems

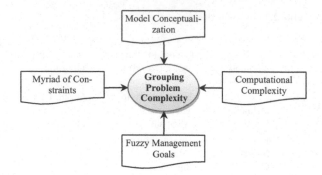

four broad categories: (1) model conceptualization or model abstraction problems,
(2) presence of myriad constraints, (3) fuzzy management goals, and (4) computational complexity.

For further understanding of the complicating features, specific challenges in
each of the four categories are discussed.

2.4 Complicating Features

In this section, various features are discussed in accordance with the categories
highlighted in the previous section.

2.4.1 Model Conceptualization

It is not always easy to conceptualize, abstract, and model a complex grouping
problem situation. The level of abstraction of the problem is affected by management goals, the cost function, the grouping phenomena, and the constraints between
the groups and within each group to be formed. In this view, problem encoding
should be structured in such a way that it represents the requisite information while
allowing for ease and efficiency of the intended iterative solution process. The
ensuing heuristic operators of the algorithm, such as selection, crossover, mutation,
and inversion, are significantly influenced by the encoding scheme, which is a
direct product of the model conceptualization and abstraction.

Most representations used in algorithms suffer from degeneracy, where multiple
chromosomes represent the same solution (Radcliffe 1991). In addition, most representations suffer from redundancy, which is the amount of excess information in
the chromosome. Degeneracy leads to inefficient exploration of the solution space
as the same configuration of groups is repeatedly explored. Therefore, a good model
conceptualization is a key component for minimization of degeneracy. This is very
crucial for an efficient algorithm.

2.4.2 Myriad of Constraints

Grouping problems in various disciplines are usually inundated with a myriad of constraints. In most cases, not all possible groupings are permitted; group formations must satisfy a number of constraints. In the presence of multiple constraints, it is highly difficult to model, let alone to solve the problems in polynomial time. The effect of such situations is that the problem becomes too restricted and computationally complex to handle. Constraints come in two basic forms as follows:

1. Hard constraints, which should be always be satisfied, e.g., a worker capacity constraints, vehicle loading constraints, or statutory regulations and
2. Soft constraints, which may be violated, but at a cost (or pseudo-cost), for example, violation of preferences and wishes.

Soft constraints can be imprecise (or fuzzy) in nature. By learning from the case studies, the identified constraints can be discussed in the following perspectives.

2.4.2.1 Intra-Group Relationship

Intra-group-related constraints are concerned with restrictions on the sequencing or the order of items in a group. If problem encoding is order dependent, the order of the items will affect the fitness of the candidate solutions (Kashan et al. 2015). For example, the sequence in a group may depict the order in which customers are visited, or the order by which tasks are executed by a worker. Therefore, the dependency between group members or items will influence the complexity of the problem. However, for an order independent grouping problem, the order does not influence constraints and fitness of the candidate solutions; for example, the team formation problem may not be order dependent. Therefore, the presence of these intra-group constraints will always influence the complexity of the problem significantly.

2.4.2.2 Inter-Group Relationship

The relationship between groups may lead to complex restrictions, which adds to more complexity of the grouping problem. For instance, in team formation, it may be required that two members in two different groups should occasionally cooperate in accomplishing a task, or should share their expertise. On the other hand, due to some working conditions, some workers may not be allowed to work in the same group, adding to the restrictiveness of the problem. In the presence of such constraints, the model becomes too restrictive and can prolong the solution process.

2.4.2.3 Group Size Limits

The limitation on the allowable number of members in a group is another crucial piece of information that influences the complexity of grouping problems. In this respect, grouping problems can have uniform or non-uniform groups. Uniform groups tend to be easier to handle than non-uniform groups, since non-uniformity will result in more constraints and influence the fitness of the candidate solutions. For example, the nonidentical parallel machines scheduling problem consist of machines with different operational characteristics, and the group of jobs that may be assigned to each machine will differ. Therefore, the group size limits will influence the complexity of the grouping problem.

2.4.2.4 Grouping Limit

This refers to the maximum allowable number of groups that may be constructed in a given problem setting. In this respect, grouping problems can be classified as constant grouping problems and variable grouping problems (e.g., bin packing problem). The number of groups may be limited to problem-specific reasons, for example, due to limited number of available healthcare workers and the number of available vehicles. In such cases, the grouping problem becomes more restricted and complex.

2.4.3 Fuzzy Management Goals

Decision makers in the modern dynamic business world make decisions under fuzziness or imprecision. In attempting to make optimal decisions, it is often realized that goals, constraints, and aspirations tend to be imprecise. This is worsened by the fact that most of the available information and problem parameters are difficult to determine precisely. Such imprecise information usually arises from three sources, namely:

1. Fuzzy wishes and expectations of the customer in regard to the desired time windows, service delivery, and service quality;
2. Fuzzy preferences worker, e.g., in terms of schedules, tasks assigned, and the related choices; and
3. Fuzzy management goals and targets upon which management aspirations are built.

For example, in vehicle routing, customers' time windows for delivery may be expressed in an imprecise manner, which may be difficult to model mathematically, let alone to solve. The same applies to home healthcare service providers.

In addition, workers in such environments may be allowed to express their desires in regard to their work schedules.

To improve service delivery and service quality, it is important to satisfy fuzzy desires and management goals to the highest degree possible. Management goals can be expressed in terms of aspiration levels and evaluated using fuzzy evaluation techniques. In the presence of fuzzy multiple goals and constraints, multi-criteria evaluation methods may be used.

2.4.4 Computational Complexity

Computational complexity arises from the curse of dimensionality as the problem size increases; grouping problems are highly combinatorial in nature. The presence of a myriad of constraints adds to the complexity of the problems. The application of classical metaheuristic algorithms such as genetic algorithms, particle swarm optimization, and the related evolutionary algorithms, their operators, and representation schemes tend to be highly redundant. This is due to the fact that the operators tend to be object-oriented rather than group-oriented, which often results in reckless breakup of the building blocks that were supposed to be promoted and improved.

Falkenauer (1994) suggested, in the case grouping problems, the representations, and resulting genetic operators should be designed in a way that will allow the propagation of groupings of objects, rather than objects, since the groupings are the inherent building blocks of the problem. It can also be argued that working on the particular positions of any one object on its own also adds to the combinatorial complexity of the problem. In view of the above, the following add to the complexity of grouping problems:

1. The presence of a myriad of constraints and variables which makes the problem highly combinatorial;
2. The need to maintain group structure and therefore prevent loss of key information;
3. The need for repair mechanisms whenever the group structure is disrupted during metaheuristic operations; and
4. The need for problem-specific constructive heuristics.

Effective techniques and heuristics should be built into metaheuristic algorithms in order to enable such approaches to address the aforementioned complexities. The next section suggests the most suitable solution approaches to industrial grouping problems.

2.5   Suggested Solution Approaches

In general, combinatorial optimization problems have a finite number of feasible solutions. However, in practice, the solution process for real-world grouping problems can be time-consuming and tedious. As a result, the overall time and cost incurred in getting accurate and acceptable results can be quite significant. As problem complexity and size increase, the effectiveness and efficiency of the current methods is limited, especially when solving modern grouping problems whose complexity continues to grow.

As realized in the selected case studies in the literature, piecemeal solution approaches have been suggested on various problem instances, including manual or basic heuristic methods, mathematical programming, metaheuristic methods, and artificial intelligence. In view of the inadequacies and shortcomings mentioned in the past approaches, the use of fuzzy multi-criteria grouping metaheuristic is highly recommended.

Intelligent fuzzy multi-criteria approaches should make use of techniques such fuzzy theory, fuzzy logic, multi-criteria decision making, and artificial intelligences. By so doing, fuzzy expert knowledge, fuzzy intuitions, fuzzy goals, and preferences can be incorporated conveniently into the modeling process.

2.6   Summary

Due to a number of complicating features, industrial grouping problems are generally NP-hard and computationally difficult to comprehend and model. Based on the recent case studies, this chapter identified characteristic complicating features that pose challenges to decision makers concerned with grouping problems. These features were classified into model abstraction, presence of multiple constraints, fuzzy management goals, and computational complexity.

Results of an in-depth taxonomic study of 18 case studies in the literature revealed a number of complicating features within the four categories. Among the methods that have been applied in these case studies, genetic algorithm is the most widely used. This indicated the great potential of the algorithm to solve a wide range of grouping problems. Realizing the inadequacies of solution methods applied, the study suggested the use of flexible, fuzzy multi-criteria grouping algorithms that hybridize fuzzy theory, fuzzy logic, grouping genetic algorithms, and intelligence. It is hoped that advances and applications of grouping genetic algorithm based on these techniques will yield remarkable progress in developing decision support tools for industrial grouping problems.

References

Agustın-Blas LE, Salcedo-Sanz S, Ortiz-Garcıa EG, Portilla-Figueras A, Perez-Bellido AM, Jimenez-Fernandez S (2011) Team formation based on group technology: a hybrid grouping genetic algorithm approach. Comput Oper Res 38:484–495

Akjiratikarl C, Yenradee P, Drake PR (2007) PSO-based algorithm for home care worker scheduling in the UK. Comput Ind Eng 53:559–583

Arogundade OT, Akinwale AT, Aweda OM (2010) A genetic algorithm approach for a real-world university examination timetabling problem. Int J Comput Appl 12(5):1–4

Behesht AK, Hejazi SR, Alinaghian M (2015) The vehicle routing problem with multiple prioritized time windows: a case study. Comput Ind Eng 90:402–413

Chen JC, Wu C-C, Chen C-W, Chen K-H (2012) Flexible job shop scheduling with parallel machines using genetic algorithm and grouping genetic algorithm. Expert Syst Appl 39: 10016–10021

Chen Y, Fan Z-P, Ma J, Zeng S (2011) A hybrid grouping genetic algorithm for reviewer group construction problem. Expert Syst Appl 38:2401–2411

Cote P, Wong T, Sabourin R (2004) Application of a hybrid multi-objective evolutionary algorithm to the uncapacitated exam proximity problem. In: Burke EK, Trick M (eds) Practice and theory of timetabling V, 5th international conference, PATAT, Pittsburgh, 18–20 August. Springer, Berlin, pp 294–312

D'Souza B, Simpson WT (2003) A genetic algorithm based method for product family design optimization. Eng Optim 35(1):1–18

Falkenauer E (1994) A New Representation and Operators for Genetic Algorithms Applied to Grouping Problems. Evol Comput 2:123–144

Falkenauer E (1996) A hybrid grouping genetic algorithm for bin packing. J Heuristics 2:5–30

Falkenauer E (1998) Genetic algorithms and grouping problems. Wiley, New York

Gendreau M, Laporte G, Musaragany C, Taillard ED (1999) A tabu search heuristic for the heterogeneous fleet vehicle routing problem. Comput Oper Res 26(12): 1153–1173

Gunn EA, Diallo C (2015) Optimal opportunistic indirect grouping of preventive replacements in multicomponent systems. Comput Ind Eng 90:281–291

Hindi KH, Yang H, Fleszar K (2002) An evolutionary algorithm for resource constrained project scheduling. IEEE Trans Evol Comput 6(5):512–518

Ho GTS, Ip WH, Lee CKM, Moua WL (2012) Customer grouping for better resources allocation using GA based clustering technique. Expert Syst Appl 39:1979–1987

Höglund H (2013) Estimating discretionary accruals using a grouping genetic algorithm. Expert Syst Appl 40:2366–2372

Joung Y-K, Noh SD (2014) Intelligent 3D packing using a grouping algorithm for automotive container engineering. J Comput Design Eng 1(2):140–151

Kashan AH, Akbari AA, Ostadi B (2015) Grouping evolution strategies: an effective approach for grouping problems. Appl Math Model 39(9):2703–2720

Kreng VB, Lee T-P (2004) Modular product design with grouping genetic algorithm—a case study. Comput Ind Eng 46:443–460

Li F, Sun Y, Ma L, Mathew J (2011a) A grouping model for distributed pipeline assets maintenance decision. In: International conference on quality, reliability, risk, maintenance, and safety engineering (ICQR2MSE), pp 601–606

Matusiak M, Koster R, Kroon L, Saarinen J (2014) A fast simulated annealing method for batching precedence-constrained customer orders in a warehouse. Eur J Oper Res 236(3):968–977

Moghadam BF, Seyedhosseini SM (2010) A particle swarm approach to solve vehicle rout-ing problem with uncertain demand: a drug distribution case study. Int J Ind Eng Comput 1:55–66

Mutingi M, Mbohwa C (2013) Task Assignment in Home Health Care: A Fuzzy Group Genetic Algorithm Approach. The 25th Annual Conference of the Southern African Institute of Industrial Engineering 2013, Stellenbosch, South Africa, 2013, 9–11 July. p 6341

Mutingi M, Mbohwa C (2014) A fuzzy-based particle swarm optimization approach for task assignment in home healthcare. S Afr J Ind Eng 25(3):84–95

Mutingi M, Mbohwa C (2015) Nurse Scheduling: A fuzzy multi-criteria simulated metamorphosis approach, Eng Lett 23(3): 222–231

Nothegger C, Mayer A, Chwatal A, Raidl GR (2012) Solving the post enrolment course timeta-bling problem by ant colony optimization. Ann Oper Res 194(1): 325–339

Ochi LS, Vianna DS, Drummond LM, Victor AO (1998) A parallel evolutionary algorithm for the vehicle routing problem with heterogeneous fleet. Future Gener Comput Syst 14:285–292

Onwubolu GC, Mutingi M (2003) A genetic algorithm approach for the cutting stock problem. J Intell Manuf 14: 209–218

Pillay N, Banzhaf W (2010) An informed genetic algorithm for the examination timetabling problem. Appl Soft Comput 10:457–467

Pitaksringkarn L, Taylor MAP (2005) Grouping genetic algorithm in GIS: a facility location modelling. J Eastern Asia Soc Transp Stud 6:2908–2920

Polata G, Kaplan B, Bingol BN (2015) Subcontractor selection using genetic algorithm. creative construction conference (CCC2015). Procedia Eng 123:432–440

Prins C (2004) A simple and effective evolutionary algorithm for the vehicle routing problem. Comput Oper Res 31:1985–2002

Radcliffe N (1991) Equivalence class analysis of genetic algorithms. Complex Syst 5:183–205

Santiago-Mozos R, Salcedo-Sanz S, DePrado-Cumplido M, Bousono-Calzon C (2005) A two-phase heuristic evolutionary algorithm for personalizing course timetables: a case study in a Spanish University. Comput Oper Res 32(7):1761–1776

Sukstrienwong A (2012) Genetic algorithm for forming student groups based on heterogeneous grouping. In: Recent advances in information science 92–97

Part II
Grouping Genetic Algorithms

Chapter 3
Grouping Genetic Algorithms: Advances for Real-World Grouping Problems

3.1 Introduction

Since the inception of grouping genetic algorithm (GGA) in the 1990s (Falkenauer 1992), significant research activities have grown over the years. New real-world grouping problems have been identified in industry, varying from service, quasi-manufacturing, manufacturing, and other industry sectors. Interestingly, the majority of the problems were solved using GGA or other related grouping algorithms such as grouping particle swarm optimization (Mutingi and Mbohwa 2013). Remarkable extensions to the GGA procedure have been done, such as fuzzy grouping genetic algorithm (Mutingi and Mbohwa 2014), for solving complex grouping problems arising in modern industry. This has indicated the ever-growing research activities in GGA and its applications.

Real-world grouping problems continue to increase in size and complexity, as the global business environment continues to grow. The number of entities in modern business systems such as logistics systems, manufacturing systems, and healthcare systems is ever increasing. Also, the level of interaction of entities within and among these systems has grown more and more complex. Continual growth in number of entities, systems, and their level of interactions results in more complexities. As a result, grouping problems in these systems are difficult to conceptualize, model, and solve due to multiple variables, objectives, and constraints involved.

In this modern competitive business environment, decision makers desire to solve problems from a multi-criteria perspective, where a number of cost functions are considered simultaneously, from a systems' perspective. In most cases, decision makers desire to consider soft constraints of problem situations by developing computational approaches that are based on multi-criteria decision making. The aim is usually to incorporate the desires, choices, or preferences of various players that are involved in the problem situation, including management goals, customer

M. Mutingi and C. Mbohwa, *Grouping Genetic Algorithms*,
Studies in Computational Intelligence 666,
DOI 10.1007/978-3-319-44394-2_3

expectations, and worker preferences. In the presence of such imprecise or fuzzy variables, developing computational solution methods is quite complex.

As the size, complexity, and variety of grouping problems continue to grow, it is hoped that current genetic algorithms will be improved and new advanced genetic techniques will be developed to handle large-scale complex grouping problems. Due to the increasing variety of the problems, designing interactive and flexible grouping algorithms is imperative. The focus of this chapter was to present the recent advances on GGA techniques, their strengths, and weaknesses, as well as their potential areas of applications. In addition, new techniques and developments are proposed, including their strengths and potential areas of application. By the end of the chapter, the following learning outcomes are envisaged:

(1) Obtain an understanding of the general overview of the grouping genetic algorithm approach;
(2) Understand the various advanced techniques and extensions of the grouping algorithm and their strengths; and
(3) Evaluate the potential application areas of the advanced techniques of the algorithm.

The next section provides an overview of the GGA approach, its key terminologies and concepts, and its constituent genetic operators. Section 3.3 presents recent and new techniques of the algorithm. Section 3.4 discusses recent and potential application areas. Section 3.5 summarizes the chapter.

3.2 Grouping Genetic Algorithm: An Overview

The concept of GGA was born from the fact that what counts in grouping problems is the group, and the genetic operators must focus on manipulating the group, rather than the individual items (Falkenauer 1992). This is because the cost functions or objectives of the problems are largely a function of the grouping structure of the candidate solution. The basic strategy, therefore, is to develop a group encoding scheme peculiar to the target grouping problem and then adapt as much as possible to the classical genetic algorithm (GA), that is, crossover, mutation, and inversion. A good example is a team formation problem, where teams or groups of experts are to be formed from a limited number of available experts, as depicted in Fig. 3.1.

According to Falkenauer (1992, 1996), classical genetic algorithms generally do not perform well on grouping problems due to high redundancy among chromosomes and context insensitivity of the crossover operator. In view of this, Falkenauer (1996) introduced the grouping genetic algorithm for solving grouping problems more efficiently. Unlike the classical GA whose focus is on individual items, GGA focuses on groups of items upon which the fitness functions and constraints are based.

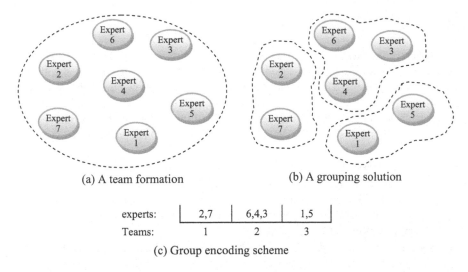

(a) A team formation   (b) A grouping solution

experts:	2,7	6,4,3	1,5
Teams:	1	2	3

(c) Group encoding scheme

Fig. 3.1 Team formation grouping problem and its group structure

3.2.1   Group Encoding

The aim of grouping (or partitioning, or clustering) problems is to construct subsets (called, *groups*) of members (called *items*) from a given set members, based on some optimization criteria. As such, the optimization or decision criteria are centered on the composition of the groups of the items. Because the decision criteria depend on the composition of the groups, the group structure has an inherent meaning to the grouping problem.

In the classical crossover operator, the value of the kth gene represents the group that the item k is in, for example, chromosome 3,221,123 encodes the grouping in which the first item is in group 3, the second in 2, the third in 2, the fourth in 1, and so on. In cases where grouping constraints exist, the crossover operator will yield many illegal groups. The group encoding scheme seeks to avoid such problems.

The encoding used for grouping problems is illustrated in Fig. 3.2, where three groups 1, 2, and 3 contain items {1, 2}, {3, 4, 5}, and {6, 7}, respectively. According to Falkenauer (1992), each group represents a gene, and the position and the order of items in a group are not insignificant. It is interesting to note that the group encoding scheme has no redundancy as in the classical GA (Falkenauer 1996; Kashan et al. 2015). The most important procedure in GGA approach is the group crossover operator.

Fig. 3.2 An illustration of the group encoding scheme for GGA approach

Items :	4,5	2,3,6	1,7
Groups:	1	2	3

3.3   Crossover

Crossover is the main genetic operator of the grouping algorithm. The operator facilitates guided information exchange between performing chromosomes with the aim for improvement. The general crossover procedure follows through six stages, as follows:

Stage 1   Two cross-points are selected randomly for both chromosomes. A crossing section is randomly selected from first parent and labeled 1 in the figure

Stage 2   The crossing section is injected into the second parent, resulting in a new offspring, which may contain repeated items (called doubles)

Stage 3   Repeated items are knocked out, avoiding the newly injected items. In the process, some of the genes of the second parent are modified; genes 4 and 6 are modified to 4' and 6'

Stage 4   The modified genes can then be adapted by way of a problem-specific heuristic, resulting in gene 7

Stage 5   Interchange the roles of the two parent chromosomes

Stage 6   Repeat stages 2 through 4 until the desired number of offspring is created

Therefore, in the actual computational implementation, the algorithm is repeated till the desired population of new offspring is obtained. To illustrate further, consider two randomly selected parent chromosomes $[1 \quad 2 \quad 3]$ containing groups of items $\{4, 5\}$, $\{2, 3, 6\}$, $\{1, 7\}$ and chromosome $[4 \quad 5 \quad 6]$ with groups of items $\{1, 5\}$, $\{2, 7\}$, $\{3, 4, 6\}$. The crossover operation for the two chromosomes is demonstrated in Fig. 3.3. The final offspring obtained are chromosomes $[4' \quad 1 \quad 5 \quad 6']$ and $[1 \quad 5 \quad 7]$.

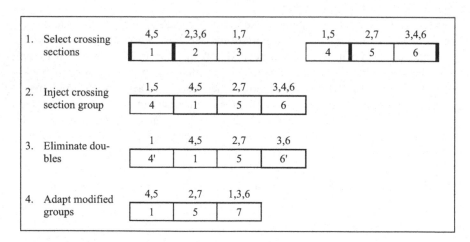

Fig. 3.3 GGA crossover operator

Following the illustration of the six states of the crossover operator, a pseudo-code of the selection and crossover operation is derived and presented as shown in Algorithm 1.

Algorithm 1 Group Selection and Crossover

1. **Input:** Population $P(t)$, *popsize*;
2. Initialize count *size* = 0;
3. **While** {*size* ≤ desired *s*} **Do**
4.   Select parents P_1 and P_2;   // select and cross two parents P_1 and P_2
5.   Select crossover sites;
6.   Inject crossing site groups; generate offspring;
7.   **If** {any *doubles* or *misses*} **Then**
8.   Eliminate doubles;
9.   Eliminate misses;
10.   **End**
11.   Adapt modified groups, if necessary;
12.   *size* = *size* + 1;   // increment offspring count
13. **End**
14. **Return**

It is easy to realize that the group crossover operator seeks to promote inheritance of groups from parent chromosomes. However, ensuring promotion of good genes is non-trivial, and developing adaptive heuristics for this purpose is quite important. The next section briefly explains the mutation operator.

3.3.1 Mutation

The grouping mutation operator must primarily work with groups, rather than items. Falkenauer (1996) proposed three general strategies, that is, (1) by creating new groups, (2) by eliminating selected existing groups, and (3) by shuffling a small number of randomly selected items among the groups. However, the exact computational implementation details will depend on the specific problem under study. It is crucial to develop innovative heuristics to intelligently guide the mutation procedure.

3.3.2 Inversion

The purpose of the inversion operator is to facilitate the transmission of good schemata from parents to offspring and ensure increased rate of sampling of the better performing schemata (Holland 1975; Falkenauer 1996). For instance, the chromosome

$$[1 \quad 2 \quad 3 \quad 4]$$

could be inverted to:

$$[1 \quad 4 \quad 3 \quad 2],$$

which increases the probability of transmitting both genes 1 and 4 together into the next generation during the next crossover operation, since they are closer together. If genes 1 and 4 are better performing ones, then good group schemata can be propagated more efficiently. However, there is a need for developing techniques for guiding the inversion operation so that best performing may have a higher chance of inversion toward promising positions in the chromosome.

3.4 Grouping Genetic Algorithms: Advances and Innovations

As discussed in the previous sections, the GGA approach utilizes a group encoded chromosomes to code candidate solutions that are then manipulated by group operators, namely group crossover, mutation, and inversion. To improve the efficiency and effectiveness of the algorithm, one should focus on innovative improvements on the GGA coding scheme and its operators. Some advances and innovations on these GGA components are presented next.

3.4.1 Group Encoding Strategies

There are several applications of GA in which grouping problems were solved, with appreciable success. However, the GGA approach has a well-structured encoding strategy that is especially designed to address grouping problems more efficiently. The original coding scheme suggested by Falkenauer (1992) encodes groups rather than items of the groups and adds that the order of items inside the groups (genes) is not relevant. Similarly, the order of the groups in the chromosome is irrelevant.

3.4.1.1 Encoding Strategy 1

This strategy is based on original scheme proposed by Falkenauer (1992), which can be explained in terms of an example. Given groups 2, 1, and 3, consisting of

Fig. 3.4 Group encoding
scheme for strategy 1

Items :	1,3,5	4,7	2,6
Group:	2	1	3

items, {1, 3, 5}, {4, 7}, and {2, 6}, respectively, the chromosomal representation
for this situation is illustrated in Fig. 3.4.

With this scheme, genetic operators are designed to work on the groups. This
strategy enables the coding to respect the spirit of the building blocks of the
algorithm, always manipulating groups as the meaningful building blocks. Each
individual item taken singly has little or no meaning to the search process
(Falkenauer 1996). However, this strategy is only suitable where there is no order
dependency between items in a group.

3.4.1.2 Encoding Strategy 2

This strategy uses a group coding scheme based on three parts, that is, the items, the
group, and the frontiers. The items part codes the items, while the group part codes
the group identity. The frontier part represents the position of last item of each
group, or the frontier of each group. Thus, the frontiers demarcate the groups. In
this case, the position of items in each group is relevant. The encoding strategy can
be explained using the previous example. Assume that we have groups 2, 1, and 3,
consisting of items, {1, 3, 5}, {4, 7}, and {2, 6}, respectively. Figure 3.5 illustrates the coding for this information. The frontiers part indicates that the digits 3, 5,
and 7 correspond to the positions of the last items of each of the three groups 2, 1,
and 3, respectively.

It is interesting to note that this coding scheme is well designed for order
dependent grouping problems such as vehicle routing problem and its variants.
Another advantage of this strategy is that genetic operators can utilize the frontiers
in their operations; for instance, mutation can be achieved by shifting a frontier one
position sideways.

3.4.2 Initialization

Initialization algorithm will always be influenced by the types of constraints in the
problem, such as inter-group constraints, intra-group constraints, group size constraints, and other problem-specific restrictions. A good initialization algorithm is
crucial for the resulting algorithm as it can significantly improve the computational

Fig. 3.5 Group encoding
scheme for strategy 2

Items:	1	3	5	4	7	2	6
Group:		2			1		3
Frontier:		3			5		7

efficiency of the algorithm. A highly diversified population of chromosomes is most preferable as this will effectively improve the evolution efficiency during the iterative search for the optimal solutions. As such, suitable initialization heuristics should be designed.

3.4.2.1 User-Generated Seeds

User-generated seeds are chromosomes that are provided by the user. For this reason, the implementation of this strategy is most appropriate when the decision maker (the user) has, through experience, some prior knowledge of good starting solutions to the grouping problem. However, it is crucial to ensure that the population has an acceptable level of diversity in order to avoid premature convergence and poor-quality solutions. Good starting solutions can enhance the iterative search process.

3.4.2.2 Random Generation

In random generation, items are assigned to groups at a probability. When grouping is order-dependent, the position of the item in the group matters. Therefore, the item is first assigned to the group and then to the position in that particular group, both at random. The process is repeated until all items are assigned. The algorithm for the random generation procedure is outlined as below, where g is the allowable number of groups, and m is the total number of available items to be assigned.

Stage 1 Input parameters: m, g, initialize $c = 0$
Stage 2 Select an item i at random, without replacement. Increment the item counter $c = c + 1$
Stage 3 Assign at random, the item i to a group j, provided the group is not full
Stage 4 Assign item i to an empty position k in group j
Stage 5 Repeat steps 1 through 4 until all items are assigned, when $c = m$

Random generation is usually suitable for grouping problems with a few hard constraints and uniform groups. As number of hard and soft constraints, constructive greedy heuristics are a viable option.

3.4.2.3 Constructive Heuristics

Constructive heuristics are problem-specific greedy heuristics that can be used to construct good initial solutions. This strategy is a good option when the grouping problem is highly constrained. In such cases, the assignment of items to groups is performed while ensuring that all hard constraints are observed. This is to ensure that only feasible solutions are generated.

To enhance the quality of the initial solutions, the heuristic should assign items to groups in an adaptive manner. For instance, this may be achieved by assigning an item i according to similarity coefficient γ_{ij}, where γ_{ij} measures the affinity between i and an unfilled group j. The algorithm for the general constructive heuristic is summarized as follows:

Stage 1 Randomly select and assign one item to each of the available g groups
Stage 2 Select any unassigned item i and assign it to an unfilled group j based on a priority of the form $p_{ij} = f(\gamma_{ij})$
Stage 3 Check feasibility based on all the hard constraints; otherwise, return to stage 2
Stage 4 Repeat steps 1 through 3 until all items are assigned to the available groups

After initialization, the generated population is evaluated and passed on to the selection and crossover operations.

3.4.3 Selection Strategies

Several selection strategies have been suggested in the literature (Goldberg 1989; Mengshoel and Goldberg 1999; Kashan et al. 2015; James et al. 2007a, b). These include deterministic sampling, remainder stochastic sampling with/without replacement, and stochastic tournament. Variants of the remainder stochastic sampling without replacement are presented. However, all the selection strategies are static; they depend on the population.

3.4.3.1 Stochastic Sampling Without Replacement

A variant of the remainder stochastic sampling without replacement is suggested. In this selection strategy, each chromosome k is selected for crossover according to its expected count e_k calculated as:

$$e_k = a \cdot f_k \Big/ \sum_{k=1}^{p} \left(\frac{f_k}{p} \right) \tag{3.1}$$

where f_k is the score function of the kth chromosome, and $a \in [0, 1]$ is an adjustment parameter. Each chromosome receives copies equal to the integer part of e_k plus additional copies obtained by treating the fractional part as a success probability.

3.4.4 Rank-Based Wheel Selection Strategy

This strategy is similar to the one suggested by James et al. (2007a, b). Individual candidates are sorted according to their cost function, and the relative position of each candidate, that is, the rank, is denoted by R_k, $k = 1, …, p$, where p is the population size. The best candidate is assigned the highest rank p, and the next best candidate is assigned rank $p - 1$, and so on. By so doing, the fitness associated with each candidate is defined by the expression:

$$f_k = \frac{2R_k}{p(p+1)} \tag{3.2}$$

The fitness values are normalized in the range, $f_k \in [0, 1]$. Note that the fitness values so obtained are then associated with the intervals of the roulette wheel.

The rank-based selection strategy is static due to the fact that the probabilities of survival depend only on the position of the individual in the ranking list. The selection process is performed with replacement, and candidates may be selected more than once.

3.4.5 Crossover Strategies

Crossover enables the GGA to explore new regions of search space, by combining the genetic information in the parents. The algorithm enables the GGA to explore new regions of the search space, a process called exploration. The crossover mechanism should be designed to avoid disruption of the group structure of the chromosomes. The choice of the crossover mechanism is influenced by the nature of the group structure of the problem, e.g., whether order-dependent or not, uniform groups or not, and other restrictions. Strategies for the crossover operation are presented.

3.4.5.1 Two-Point Group Crossover

This crossover strategy is an adaptation from the conventional two-point crossover mechanism. Figure 3.6 illustrates an example of the crossover mechanism based on two randomly selected parent chromosomes, $P_1 = \begin{bmatrix} 1 & 2 & 3 & 4 \end{bmatrix}$ corresponding to groups of items $\{1, 3\}$, $\{4, 7\}$, $\{2, 6\}$, and $\{8, 5, 9\}$, and $P_2 = \begin{bmatrix} 5 & 6 & 7 & 8 \end{bmatrix}$ corresponding to groups of items $\{8, 7, 2\}$, $\{3, 5\}$, $\{6, 9\}$, and $\{1, 4\}$.

The crossover strategy described in terms of the example above is summarized into four stages as outlined in the algorithm below:

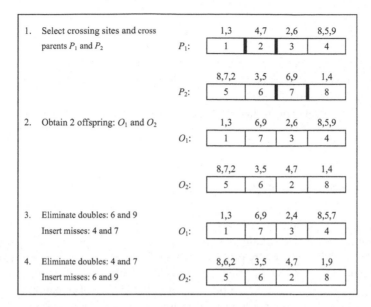

Fig. 3.6 Group crossover operation

Stage 1   Randomly selected two parent chromosomes, P_1 and P_2, and select crossing sections for the two chromosomes.

Stage 2   Cross P_1 and P_2 by interchanging the crossing sections of the parent chromosomes. Obtain two offspring O_1 and O_2, which may have repeated items (called doubles) or missing items (called misses).

Stage 3   Eliminate the doubles, using a constructive repair mechanism, avoiding the crossed items.

Stage 4   Insert the missing items into groups, beginning from where doubles were eliminated, subject to group size limits.

The crossover mechanism is repeated until the required population of new off-spring is created. After crossover, a new population, *newpop*, is created by combining the best performing offspring and the current population *oldpop*. The *newpop* is passed on to the mutation operation.

3.4.5.2 Adaptive Crossover

Adaptive crossover is based on the premise that the probability of crossover must be high in the early generations and moderate in the last ones in order to enhance a balanced approach to exploration and exploitation of the solution space. Therefore, an adaptive crossover probability is defined for computational implementation:

$$p^c(t) = p_0^c + \frac{t}{T}\left(p_0^c - p_f^c\right) \tag{3.3}$$

where $p^c(t)$ is the crossover probability at generation t; T is for the predefined maximum number of generations; and p_0^m and p_f^m are the initial and final values of the mutation probability, respectively.

Other expressions and forms of adaptive crossover can be devised in a similar manner.

3.4.6 Mutation Strategies

The role of the mutation operator is to exploit the search space in the neighborhood of the current candidate solutions, so as to improve the solutions. As opposed to crossover, mutation essentially provides GGA with a local search capability, called intensification or exploitation. This is achieved by making small perturbations on selected chromosomes, at a low probability p^m. The choice of mutation strategy is influenced by the group structure of the problem. During the search process, it is important to balance exploration and exploitation using adaptive mutation.

3.4.6.1 Swap Mutation

The swap mutation strategy is an adaptation from the mutation by swapping that is normally used in the conventional GA. However, in GGA, the swap mutation mechanism is centered on the groups rather than individual items in the conventional GA. Mutation enables small probabilistic perturbations on the population chromosomes, aimed at improving the current candidate solutions by searching in their neighborhoods. To safeguard against infeasibilities, a repair mechanism is employed whenever any hard constraint is violated during mutation. Figure 3.7 demonstrates the swap mutation mechanism based on an example of a chromosome [{1, 3}{6, 9}{2, 4}{8, 5, 7}].

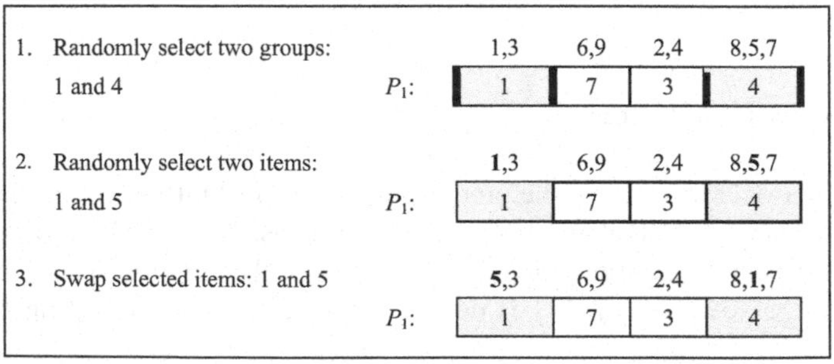

			1,3	6,9	2,4	8,5,7
1.	Randomly select two groups: 1 and 4	P_1:	1	7	3	4
			1,3	6,9	2,4	8,5,7
2.	Randomly select two items: 1 and 5	P_1:	1	7	3	4
			5,3	6,9	2,4	8,1,7
3.	Swap selected items: 1 and 5	P_1:	1	7	3	4

Fig. 3.7 Swap mutation operation

The mutation operation illustrated above can be summarized into a four-stage algorithm as follows:

Stage 1   At a probability p^m, successively select a chromosome from the current population.
Stage 2   Randomly choose two different groups from the selected chromosome.
Stage 3   Randomly select two items, one from each group of the selected groups.
Stage 4   Swap the selected items and repair the resulting chromosome if necessary.

However, the swap mutation mechanism is performed subject to problem-specific hard grouping constraints.

3.4.6.2   Split Mutation

This mutation strategy involves randomly selecting a group and splitting into two different groups. The probability q_j of selecting a group j ($j = 1, \ldots, m$) for mutation is proportional to the size of the group; the larger the relative size of a group, the higher the chances of its selection for split mutation. The logic behind this setting is that groups that are growing too large are expected to be reduced, while small groups should be expected to increase in size. This procedure encourages balanced grouping.

The split mutation mechanism can be summarized as outlined in the following algorithm:

Stage 1   Select a group j' at a probability q_j, which is proportional to its relative size
Stage 2   Reassign every item i from the selected group j' to any other group j at probability p_{ij}
Stage 3   Repair the chromosome, if necessary

Each item i from the split group j' is then probabilistically assigned to each group j at a probability p_{ij}, which is a function of the similarity coefficient γ_{ij} between item i and group j. Figure 3.8 demonstrates the split mutation procedure based on an example of chromosome $[\{1, 3\}\{5, 6\}\{2, 4, 7, 8\}]$. The mutation process split

1. Select a group with probability q_j	1,3	5,6	2,4,7,8	
Group 3 is selected	1	2	3	

2. Assign item $i = 2$ to j at	1,3	5,6	2	-
probability p_{ij}	1	2	3	3'

3. Repeat stage 2 for items 4,7	1,3	5,6	2,7	4,8
and 8	1	2	3	3'

Fig. 3.8 Mutation by splitting a group

group 3 and reassigned its items to existing groups, resulting in chromosome
[{1, 3}{5, 6}{2, 7}{4, 8}].

3.4.6.3   Merge Mutation

The merge mutation consists in concatenating two groups selected based on an
adaptive probability. As such, the probability of choosing the group is a function of
size of the group; the higher the group size, the lower the probability of being
selected for the merge mutation. Thus, the probability q_j for selecting group j for
mutation is a function of the size of the group. As in the split mutation mechanism,
the idea behind increasing the chances of merging small groups is to encourage
balanced grouping. This is quite practical in a number of problems. For instance, in
vehicle routing (Liu et al. 2009), home healthcare scheduling (Mutingi and
Mbohwa 2014), and task assignment (Muting and Mbohwa 2014), it important to
balance workload assignment among workers, where workload is a function of the
number of customers, patients or tasks assigned.

Figure 3.9 demonstrates the merge mutation operator based on an example,
using chromosome $[1 \quad 2 \quad 3 \quad 4]$ with groups of items [{1, 3}{6, 9, 2}{4}
{8, 5, 7}].

The merge mutation operator can be summarized as shown in the following
algorithm:

Stage 1   Select any two groups, j_1 and j_2, with probabilities q_{j1} and q_{j2},
          respectively
Stage 2   Merge the two groups, j_1 and j_2, into one group
Stage 3   Reassign any unassigned item at a probability p_{ij} if any hard constraint is
          violated

The resulting chromosome is [{6, 9, 2}{1, 4, 3}{8, 1, 7}], with evenly distributed group sizes. In the case where a grouping constraint is broken because of
the merge mutation process, the chromosome is restructured by reassigning each
unassigned item i to group j at a probability p_{ij}.

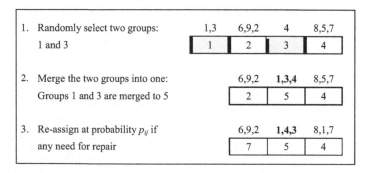

Fig. 3.9 Mutation by merging groups

3.4.6.4 Adaptive Mutation

An adaptive mutation operator assumes that the probability of mutation is smaller in the early generations than in the latter generations. The aim is to increase the chances of the algorithm to escape from the local minima in the last phases of the search process. In the adaptive mutation, procedure enables the algorithm to intensify its search process in the neighborhood of the best solutions. The following expression can be used for computing the adaptive mutation probability:

$$p^m(t) = p_0^m + \frac{t}{T}\left(p_f^m - p_0^m\right) \tag{3.4}$$

where $p^m(t)$ is the crossover probability at generation t; T is for the predefined maximum number of generations; and p_0^m and p_f^m are the initial and final values of the mutation probability, respectively. In the same way, other forms of expressions can be developed for adaptive crossover.

3.4.7 Inversion

The inversion operator is designed to enable the crossover operator to select a variety of groups in the crossover operation. This also helps to improve the chances of involving the shifted groups with other operators. This is done randomly, at a low probability p^i, or with a purpose-driven repositioning of groups and/or items. Groups which are close together in the sequence have a higher probability of being shifted together.

3.4.7.1 Two-Point Inversion

The two-point inversion overturns items within two inversion sites, selected at random. If the resulting chromosome violates any problem-specific constraints, a repair mechanism can be utilized. Figure 3.10 illustrates the two-point inversion procedure, based on an example of the chromosome $[1 \quad 2 \quad 3 \quad 4]$ with groups of items $[\{1, 3\}\{4, 7\}\{2, 6\}\{8, 5\}]$.

The inversion mechanism overturns groups 2 and 3 lying between the two inversion sites. As a result, the chromosome $[\{1, 3\}\{4, 7\}\{2, 6\}\{8, 5\}]$ is transformed to $[\{1, 3\}\{2, 6\}\{4, 7\}\{8, 5\}]$. In other words, the chromosome $[1 \quad 2 \quad 3 \quad 4]$ is transformed to $[1 \quad 3 \quad 2 \quad 4]$, where all the groups within the inversion sites are rewritten in the reverse order.

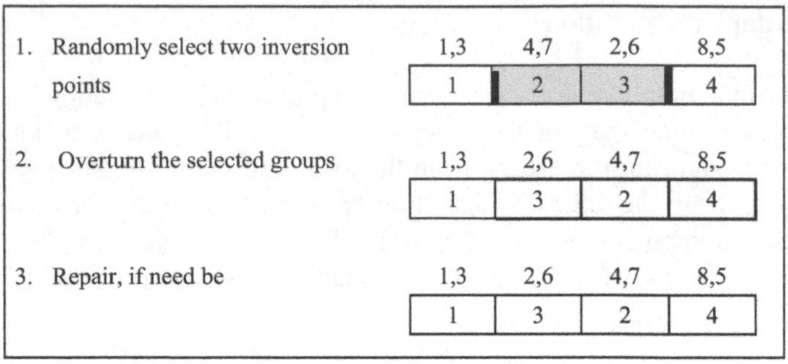

1. Randomly select two inversion points	1,3	4,7	2,6	8,5
	1	2	3	4
2. Overturn the selected groups	1,3	2,6	4,7	8,5
	1	3	2	4
3. Repair, if need be	1,3	2,6	4,7	8,5
	1	3	2	4

Fig. 3.10 Two-point inversion operator

3.4.7.2   Single-Point Inversion

The single-point inversion is derived from the single-point mutation of the conventional GA. The mechanism selects any two different groups at random and interchanges the groups together with the items. Again, if the resulting chromosome violates any constraints, a repair mechanism is employed if need be. Figure 3.11 shows the single-point inversion procedure, based on an example of the chromosome $[1 \quad 2 \quad 3 \quad 4]$ containing groups of items {1, 3}, {4, 7}, {2, 6}, and {8, 5}. The single-point inversion mechanism transforms the chromosome to $[3 \quad 2 \quad 1 \quad 4]$ with corresponding groups of items {2, 6}, {4, 7}, {1, 3}, and {8, 5}.

Like the two-point inversion, the single-point inversion is static and thus is not adaptive to the dynamic changes in the search process. An adaptive inversion mechanism is suggested.

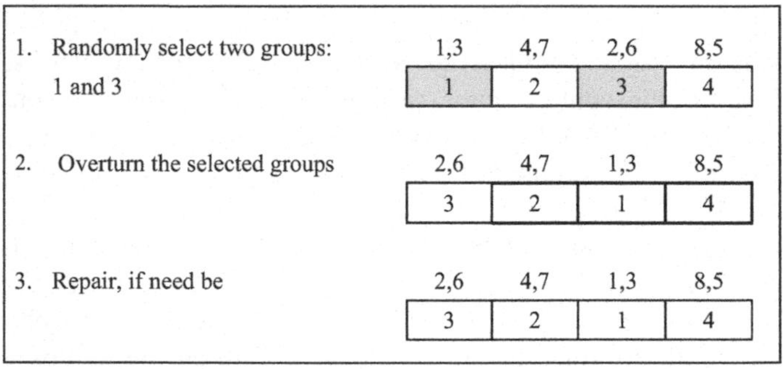

1. Randomly select two groups: 1 and 3	1,3	4,7	2,6	8,5
	1	2	3	4
2. Overturn the selected groups	2,6	4,7	1,3	8,5
	3	2	1	4
3. Repair, if need be	2,6	4,7	1,3	8,5
	3	2	1	4

Fig. 3.11 Single-point inversion operator

3.4.7.3   Adaptive Inversion

For a more adaptive inversion, the inversion operator should be performed at a probability p^i which changes with iteration count and an adjustment factor $\alpha \in [0, 1]$. The adjustment factor can be formulated as a parameter that changes according to the state of the search process. Therefore, the probability p^i can be expressed as a dynamically decreasing function as follow:

$$p^i(t) = p_0^i e^{-\alpha(t/T)} \tag{3.5}$$

where α is a constant in the range [0, 1]; t is the iteration count; T is the maximum count; and p_0 is the initial inversion probability.

3.4.8   Replacement Strategies

As the algorithm progresses from one generation to the next, the population at a generation $t + 1$ is obtained by replacing the chromosomes in the current population j through the application of the grouping genetic operators (crossover, mutation, and inversion).

In order to ensure that GGA does not lose best performing candidates, an elitist strategy should be applied, where the best candidates in generation j are automatically advanced onto generation $t + 1$. This strategy is generally beneficial as it ensures that the best candidate encountered so far is always kept by the algorithm. This also helps to reduce the computational complexity. However, this may not always workout for the best of the results.

Completely disregarding poor candidates in the current population is not always the best way to go since that can result in premature convergence as less fit candidates are culled with no opportunity to evolve into something better. For instance, there may exist in the current population some less performing candidates that could result in excellent ones by slight genetic alterations. As such, keeping other less performing candidates may be beneficial in the next iterations ensuring that the best solution. A cautious trade-off between the best and less performing candidates is essential. To improve diversity in the population and to retain potential candidates in the population, probabilistic crowding (Lianga and Leung 2011) may be useful. With this strategy, individual candidates with low fitness are probabilistically conserved and advanced into the next generation.

3.4.9 Termination Strategies

The termination of the GGA iterative search process should continue until the search space is sufficiently explored and exploited. Therefore, the termination criteria should be constructed accordingly. Some of the suggested criteria are as presented.

3.4.9.1 Iteration Count (*ItCount*)

This is the most common criterion, where the algorithm is designed to terminate whenever the predetermined iteration count *maxItCount* is reached. It is essential to choose a sufficiently large *maxItCount* to ensure sufficient search for near-optimal solutions. However, it is difficult to determine the right maximum iteration count, *maxItCount*.

3.4.9.2 Iterations Without Improvement (*ItWithoutImp*)

In this strategy, the termination condition is met when a prespecified number of iterations *maxItWithoutImp* are reached with no significant improvement in the fitness of the current best solution. However, a hybrid approach may be more realistic as it seeks for a balanced trade-off between two and more criteria.

3.4.9.3 Hybrid Criteria

An adaptive approach to termination, developed from a combination of the criteria above, is preferred and recommended for grouping problems. This allows the algorithm to search the solution space sufficiently enough by dynamically considering the current state of the search process and the current solutions. In this case, the termination condition is more flexible and adaptive.

Algorithm 2 Termination condition

1. **Input**: itCount, itWithoutImp, maxItCount, α, β
2. Terminate = false   //initialize terminate;
3. Compute α = f (current conditions)   //adjust parameters α, β ;
4. Compute β = f (current conditions);
5. **If** (itCount $\geq \alpha \cdot$ maxItCount & itWithoutImp $\geq \beta \cdot$ maxItWithoutImp) **Then**
6.   terminate = true;
7. **End**
8. **Return:** terminate

Parameters α, β are adjusted according to the current state of the search process. The values of the parameters can be used to control further exploration and exploitation of the solution space.

The next section presents recent and potential areas of application where further research is most likely to be fruitful.

3.5   Application Areas

Though the GGA approach has been applied to a number of industrial grouping problems in the literature, it can be seen that more grouping problems can be modeled and addressed using this approach. Some of the application areas in industry are healthcare, tertiary education, warehouse and logistics, manufacturing systems, design, maintenance management, and business economics, as well as human resource management.

Table 3.1 presents an outline of brief descriptions of some of the most recent active research activities on grouping problems in various sectors of industry.

Table 3.1 Recent grouping problem areas for further research

Application area	Brief description	References
1. Healthcare	Care task assignment in healthcare	Mutingi and Mbohwa (2014)
	Handicapped person transportation problem	Rekiek et al. (1999)
	Home healthcare staff scheduling	Mutingi and Mbohwa (2014)
2. Information technology	Wi-fi network deployment problem	Agustín-Blas et al. (2011)
	Multiprocessor scheduling	Singh et al. (2009)
3. Tertiary education	Reviewer group construction problem	Chen et al. (2011a, b)
	Student/learners grouping	Chen et al. (2012a, b)
	Exam timetabling	Rakesh et al. (2014)
4. Maintenance	Preventive maintenance planning for multi-component systems	Gunn and Diallo (2015), De Jonge et l. (2016)
	Group maintenance scheduling for network maintenance	Li et al. (2013)
5. Warehouse and logistics	Manual order batching in warehouse distribution	Henn and Wascher (2012)
	Fleet size and mix vehicle routing	Liu et al. (2009)
	Container loading	Joung and Noh (2014)
	Bin packing	Kaaouache and Bouamama (2015)

(continued)

Table 3.1 (continued)

Application area	Brief description	References
6. Economics and business	Estimation of discretionary accruals	Höglund (2013)
	Customer grouping for better resources allocation	Ho et al. (2012)
	Economies of scale	Falkenauer (1998)
	Selection of audit teams	Caetano et al. (2013)
7. Manufacturing systems	Cellular manufacturing system design	Filho and Tiberti (2006a, b)
	Flexible job shop scheduling	Chen et al. (2012)
	Assembly line balancing	Kaaouache and Bouamama (2015)
	Team formation for proper resource management	Wi et al. (2009), Strnad and Guid (2010)
8. Design	Modular design for green life cycle engineering	Chen and Martinez (2012)
	Product family design optimization	D'Souza and Simpson (2002)

Further research on the prospective application of grouping genetic algorithms in these areas looks promising and interesting.

It will be interesting to pursue further research on some or all of the listed application areas. The next section summarizes the work presented in this chapter.

3.6   Summary

Since its inception in the 1990s, the GGA approach has been instrumental in solving a number of industrial grouping problems. Over the years, a significant amount of research brought up new techniques and new areas of applications of the algorithm. However, further research on the algorithm and its potential applications continue to grow. This chapter focused on recent advances on GGA techniques and their potential application areas. In addition, new techniques and developments were proposed and presented, including their potential advantages and their potential areas of applications.

It will be interesting to experiment on some of the recent and proposed GGA techniques. This will be dealt with in the forthcoming sections. The next section of this book focuses on illustrative applications of these techniques.

References

Agustın-Blas, LE, Salcedo-Sanz S, Ortiz-Garcıa EG, Portilla-Figueras A, Pérez-Bellido AM, JimJiménez-Fernández S (2011) Team formation based on group technology: a hybrid grouping genetic algorithm approach. Comput Oper Res 38: 484–495

Caetano SS, Ferreira DJ, Camilo CG (2013) Multi-objective genetic algorithm for competency-based selection of auditing teams. J Softw Syst Dev Article ID 369217:1–13. doi:10.5171/2013.369217

Chen AL, Martinez DH (2012) A heuristic method based on genetic algorithm for the baseline-product design. Expert Syst Appl 39(5):5829–5837

Chen JC, Wu C-C, Chen C-W, Chen K-H (2012a) Flexible job shop scheduling with parallel machines using Genetic Algorithm and Grouping Genetic Algorithm. Expert Syst Appl 39 (2012):10016–10021

Chen R-C, Chen S-Y, Fan J-Y, Chen Y-T (2012) Grouping partners for cooperative learning using genetic algorithm and social network analysis. In: The 2012 international workshop on information and electronics engineering (IWIEE). Procedia Engineering, vol 29, pp. 3888–3893

Chen Y, Fan Z-P, Ma J, Zeng S (2011) A hybrid grouping genetic algorithm for reviewer group construction problem. Expert Syst Appl 38:2401–2411

Chen JC, Wu C-C, Chen C-W, Chen K-H (2012) Flexible job shop scheduling with parallel machines using genetic algorithm and grouping genetic algorithm. Expert Syst Appl 39 (2012):10016–10021

D'souza B, Simpson TW (2002) A genetic algorithm based method for product family design optimization (2003). Eng Optim 35(1):1–18

de Jonge B, Klingenberg W, Teunter R, Tinga T (2016) Reducing costs by clustering maintenance activities for multiple critical units. Reliab Eng Syst Saf 145:93–103

Falkenauer E (1992) The grouping genetic algorithms—widening the scope of the GAs. Belgian J Oper Res, Stat Comput Sci 33:79–102

Falkenauer E (1996) A hybrid grouping genetic algorithm for bin packing. J Heuristics 2:5–30

Falkenauer E (1998) Genetic algorithms for grouping problems. Wiley, New York

Filho EVG, Tiberti AJ (2006a) A group genetic algorithm for the machine cell formation problem. Int J Prod Econ 102:1–21

Filho EVG, Tiberti AJ (2006b) A group genetic algorithm for the machine cell formation problem. Int J Prod Econ 102:1–21

Gunn EA, Diallo C (2015) Optimal opportunistic indirect grouping of preventive replacements in multicomponent systems. Comput Ind Eng 90(2015):281–291

Goldberg DE (1989) Genetic Algorithms: Search, Optimization and Machine Learning, Reading, MA: Addison Wesley

Henn S, Koch S, Wäscher G (2012) Order batching in order picking warehouses: a survey of solution approaches. In: Manzini R (ed) Warehousing in the global supply chain: Advanced models, tools and applications for storage systems. Springer, London, pp 105–137

Ho GTS, Ip WH, Lee CKM, Moua WL (2012) Customer grouping for better resources allocation using GA based clustering technique. Expert Syst Appl 39:1979–1987

Höglund H (2013) Estimating discretionary accruals using a grouping genetic algorithm. Expert Syst Appl 40:2366–2372

Holland John H (1975) Adaptation in natural and artifical systems. University of Michigan Press, Ann Arbor, MI

James T, Vroblefski M, Nottingham Q (2007a) A hybrid grouping genetic algorithm for the registration area planning problem. Comput Commun 30(10):2180–2190

James TL, Brown EC, Keeling KB (2007b) A hybrid grouping genetic algorithm for the cell formation problem. Comput Oper Res 34:2059–2079

Joung Y-K, Noh SD (2014) Intelligent 3D packing using a grouping algorithm for automotive container engineering. J Comput Des Eng 1(2):140–151

Kaaouache MA, Bouamama S (2015) Solving bin packing problem with a hybrid genetic algorithm for VM placement in cloud. Proc Comput Sci 60:1061–1069

Kashan AH, Akbari AA, Ostadi B (2015) Grouping evolution strategies: an effective approach for grouping problems. Appl Math Model 39(9):2703–2720

Li F, Ma L, Sun Yong, Mathew J (2013) Group maintenance scheduling: a case study for a pipeline network. In Engineering asset management 2011: proceedings of the sixth annual world congress on engineering asset management (Lecture notes in mechanical engineering), Springer, Duke Energy Center, Cincinnati, Ohio, pp. 163–177

Lianga Y, Leung K-S (2011) Genetic Algorithm with adaptive elitist-population strategies for multimodal function optimization. Appl Soft Comput 11:2017–2034

Liu S, Huang W, Ma H (2009) An effective genetic algorithm for the fleet size and mix vehicle routing problems. Transp Res Part E 45:434–445

Mengshoel OJ, Goldberg DE (1999) Probability crowding: deterministic crowding with probabilistic replacement. In: Banzhaf W (ed) Proceedings of the international conference. GECCO-1999, Orlando, FL, 1999, pp. 409–416

Mutingi M, Mbohwa C (2013) Home healthcare worker scheduling: a group genetic algorithm approach. Proceedings of the world congress on engineering 2013, UK, 3–5 July, 2013, London, UK, pp. 721–725

Mutingi M, Mbohwa C (2014) A fuzzy-based particle swarm optimization approach for task assignment in home healthcare. S Afr J Ind Eng 25(3):84–95

Rakesh P, Badoni DK, Gupta Pallavi Mishra (2014) A new hybrid algorithm for university course timetabling problem using events based on groupings of students. Comput Ind Eng 78:12–25

Rekiek B, De Lit P, Pellichero F, Falkenauer E, Delchambre A (1999) Applying the equal piles problem to balance assembly lines. Proceedings of the 1999 IEEE international symposium on assembly and task planning Porto, Portugal, July 1999

Singh A, Sevaux M, Rossi A (2009) A hybrid grouping genetic algorithm for multiprocessor scheduling. In Contemporary computing 40, communications in computer and information science. Springer Verlag Berlin Heidelberg, pp. 1–7

Strnad D, Guid N (2010) A fuzzy-genetic decision support system for project team formation. Appl Soft Computing: 1178–1187

Wi H, Oh S, Mun J, Jung M (2009) A team formation model based on knowledge and collaboration. Expert Syst Appl 36(5):9121–9134

Chapter 4
Fuzzy Grouping Genetic Algorithms: Advances for Real-World Grouping Problems

4.1 Introduction

Grouping genetic algorithm (GGA) is a metaheuristic algorithm for solving grouping (clustering or partitioning) problems such as bin packing, vehicle routing, and economies of scale. Originally developed by Falkenauer (1992, 1994, 1996), the algorithm is a development from the well-known classical genetic algorithm (GA), with unique grouping genetic operators, that is, crossover, mutation, and inversion. Like the classical GA, the GGA mechanism is inspired by the biological mechanisms of evolution, survival of the fittest, and heredity, coupled with human reasoning to achieve an efficient and effective iterative search and optimization procedure for grouping problems. Given a set of candidate solution (called population) coded as chromosomes and a specific function to be optimized, a selection mechanism is used to pick potential chromosomes (called parents) for creating the next generation through a crossover operator. Subsequently, two parents are crossed to produce two new chromosomes (called offspring). Further, each resulting chromosome is subjected to a mutation operator which randomly mutates a few genes of in each chromosome. Last but not least, the inversion operator shifts the position of some genes in a chromosome by overturning the genes between two randomly selected crossing points in order to improve the crossover operation in the next generation. Based on their fitness, the offspring then compete with the old chromosomes for a place in the next generation. Typically, fitter chromosomes have higher chances of survival into the next generation than the weaker ones. The process is repeated iteratively until a termination criterion is met, usually when a satisfactory solution is obtained.

In the real-world decision making, various kinds of uncertainties, such as expert information, qualitative statements, vagueness, and imprecision, are often involved. Further to that, most of the management goals are seldom expressed precisely. With multiple imprecise goals, global optimization becomes even more complex. Owing to GGA's inherent stochastic nature, common problems arise in the global

© Springer International Publishing Switzerland 2017
M. Mutingi and C. Mbohwa, *Grouping Genetic Algorithms*,
Studies in Computational Intelligence 666,
DOI 10.1007/978-3-319-44394-2_4

optimizing process and the convergence speed. Improving the search and optimization capability and the convergence of chromosomes are two crucial issues to
be addressed in GGA. These issues have also been raised by several authors for the
case of GA (Herrera and Lozano 2003; Chao et al. 1999; Hu et al. 2004, 2008;
Eiben and Smith 2003; Hu and Wu 2007; Arnone et al. 1994).

As is the case with GAs, there are many parameters which may affect the
optimizing capability and convergence speed of the GGA approach. These include
the evaluation, crossover, mutation, and inversion probabilities. A proper
fine-tuning and control of these parameters will ensure satisfactory performance of
the GGA mechanism. A fuzzy adaptive control approach is quite promising.

Since its inception in the 1990s, the algorithm has gained much attention in the
research community. Though the algorithm has been widely used in global optimization, with remarkable success, researchers in the area have faced challenges
relating to its computational issues and the ultimate performance. Some of the
issues are outlined as follows:

1. The GGA is influenced by several factors and genetic parameters affecting its
   ultimate performance;
2. The interactive relationship between the algorithm's performance and these
   factors and parameters is complex;
3. The appropriate values of the interacting factors and parameters are difficult to
   determine in a precise manner.

In view of the above, the use of fuzzy mechanism is the most viable way to
model the complex interactions and relationships between the factors and the
parameters. Therefore, GGAs require expert supervision during their iterative
process, for the following reasons: (i) to detect the emergence of a good solution,
(ii) to fine-tune and adapt the algorithm parameters, and (iii) to monitor the evolution process so as to prevent undesirable behavior such as premature convergence. In practice, the general behavior of GGA and the interrelations between its
genetic operators are very complex. The use of fuzzy control theory in algorithm
development may yield improved results. In the presence of many possible inputs
and outputs for the current fuzzy control systems, fuzzy rule bases are not readily
available for specific problems because developing good fuzzy rule bases is a
complex task (Herrera and Lozano 2003). Fine-tuning, control, and adaptation of
the behavior of genetic parameters and the overall GGA are difficult to describe in a
precise manner.

This chapter focuses on the advances and innovations in the use of fuzzy
evaluation and fuzzy logic control concepts and their infusion in the GGA mechanism. The specific objectives are as follows:

1. To evaluate the complexities associated with the genetic parameters of the
   genetic operators;
2. To formulate fuzzy-based computational strategies for improving the search and
   optimization process of the GGA mechanism; and

3. To incorporate fuzzy-based strategies into the genetic operators of the grouping algorithm.

In developing the fuzzy-based GGA, the grouping genetic operators are enriched with fuzzy logic control and other fuzzy theoretic concepts to enable the GGA to accommodate expert opinion and guidance during its search and optimization process, and to handle complex real-world grouping problems with fuzzy characteristics. It is hoped that the proposed fuzzy GGA (FGGA) presented in this chapter is an effective and efficient algorithm for solving real-world grouping problems, even in a fuzzy environment. Avenues for further research in this direction are suggested.

The rest of the chapter is structured as follows: The next section gives a background to fuzzy logic control. Details to the advances and innovations in fuzzy grouping genetic algorithms are given in Sect. 4.3. Section 4.4 outlines the potential application areas for the grouping algorithm. Section 4.5 summarizes the chapter and suggests further research avenues.

4.2 Preliminaries: Fuzzy Logic Control

Fuzzy logic control (FLC) is a useful technique to enhance the performance of adaptive global metaheuristic optimization algorithms (Cordon et al. 2001), such as simulated metamorphosis (Mutingi and Mbohwa 2015), fuzzy simulated evolution algorithm (Li and Kwan, 2001), group particle swarm optimization (Mutingi and Mbohwa 2014a, b, c), and other evolutionary algorithms (Mutingi and Mbohwa 2014a, b).

Fuzzy logic control concepts are quite handy when the parameters and their level of interaction are imprecise and complex, so much that they cannot be modeled using conventional means. In particular, fuzzy logic control is very useful as it takes into account the imprecise nature of the expert opinion on the values of the parameters, their associated interactions, and how they influence the performance of the algorithm. The most essential part of the fuzzy-based control is a set of fuzzy rules, that is, fuzzy conditional statements of the form,

If (antecedent is satisfied) **Then** Consequent

which basically consists of fuzzy statements. A fuzzy logic system must have a knowledge base containing information provided by the expert in the form of linguistic control fuzzy rules. More information on fuzzy logic control can be obtained in Driankow (1993) and Cordon et al. (2001).

One of the worst challenges of global optimization and its computation is the adaptation of the optimization search parameters, such as genetic control parameters in genetic algorithms. However, the use of fuzzy logic control (FLC) or any fuzzy-based control is quite beneficial. In a fuzzy-based global optimization, a rule base can be used to capture and represent a range of adaptive strategies for the global optimization algorithm. Figure 4.1 shows the proposed fuzzy global

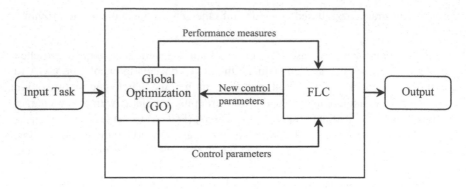

Fig. 4.1 A fuzzy global optimization model

optimization model. First, the FLC accepts a combination of inputs in the form of performance measures or current control parameters, and then generates outputs as new control parameters. In turn, the current performance measures of the global optimizer are, again, sent to the FLC, which then generates new control parameters for the next generation. The iterative process continues in that manner until a termination criterion is satisfied.

4.3 Fuzzy Grouping Genetic Algorithms: Advances and Innovations

Like classical genetic algorithms, the behavior of the grouping genetic algorithms (GGA) is heavily influenced by the trade-off between *exploration* and *exploitation*. Exploration seeks to reach out for new solutions in the unvisited regions of the search space, while exploitation is the intensified search for improved solutions in the neighborhood of the current solutions that were previously obtained by exploration. To obtain the best computational results, genetic parameters, such as diversity measure, convergence measure, crossover probability, and mutation probability, have to be controlled to cautiously balance exploration and exploitation. Basically, fuzzy GGA is a grouping genetic algorithm that utilizes one or more fuzzy controlled genetic parameters to improve the search and optimization process of the algorithm.

Figure 4.2 shows the general flowchart for the proposed fuzzy grouping algorithm. First, the algorithm accepts input in the form of genetic parameters, such as crossover, mutation, and inversion probabilities. An initial population is generated randomly or through constructive heuristics and then evaluated for fitness. Based on the current performance measures such as maximum fitness value, average fitness value, and genetic control parameters (crossover, mutation, and inversion probabilities), new adjusted parameters are then returned for use in the next generation.

Fig. 4.2 Fuzzy grouping
genetic algorithm approach

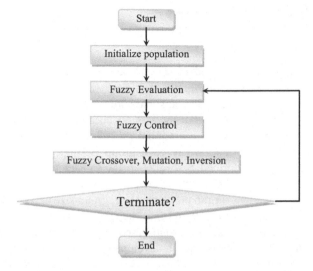

The iterative loop continues till the user-defined termination condition is satisfied. The pseudo-code of the proposed FGGA procedure is presented in Algorithm 1, as follows:

Algorithm 1 Fuzzy grouping genetic algorithm

1. **Input**: initial p^c, p^m, and p^i, population size p;
2. Initialize population $P(0)$
3. **Repeat**
4.   Fuzzy evaluation (P);
5.   Fuzzy control (P);   // classification according to the fitness
6.   Obtain fuzzy p_n^c p_n^m and p_n^i;
7.   Fuzzy crossover (), mutation (), inversion ();
8. **Until** (termination criteria is fulfilled)
9. **Return**: P

The specific elements of the proposed FGGA, including chromosome representation, initialization, fuzzy fitness evaluation, crossover, mutation, and inversion, are presented in this section.

4.3.1 FGGA Coding Scheme

Oftentimes, the coding scheme for FGGA is similar to the one for GGA. The primary aim is to ensure that the group structure of the problem is represented. The group-based coding should ensure that relevant information can be decoded

Fig. 4.3 FGGA chromosome
coding scheme

Learners: | 1, 3, 5, 12 | 4, 7, 9, 10 | 2, 6, 8, 11 |

Group: 1   2   3

efficiently when calculating objective functions. In addition, the coding should ensure ease of information decoding when testing constraints. Some schemes go a step further to ensure that some hard constraints are directly infused into the coding structure so much that there is no need to check for violation of the constraints. A good example is a grouping problem where a class of 12 learners is to be split into groups of 4, where each group is supposed to carry out a specific project. Groups are formed according to their characteristic preferences. Since the group size is fixed, the length of the chromosome is also fixed. Figure 4.3 represents the group coding for the problem. Here, group size and class size (chromosome length) constraints are hardwired into the coding structure.

Deriving from the group-based criterion, and the specific problem context, an initialization procedure is developed.

4.3.2 Initialization

A number of methods are available for the creation of the initial population, including (i) random assignment of items (e.g., learners) to groups, (ii) seeds or solutions known a priori by the expert user, (iii) greedy heuristics which are usually problem dependent, or (iv) a combination of the above. The initialization procedure should typically create a population with good feasible solutions and an acceptable level of diversity.

4.3.3 Fuzzy Fitness Evaluation

In fuzzy global optimization, the choice of the evaluation approach will influence the quality measures that are used to express the fitness of population chromosomes. This, in turn, will determine the performance measures that drive the whole search process. Two evaluation approaches are presented.

4.3.3.1 Multifactor Evaluation

In, the real-world settings, grouping problems are characterized with multiple criteria. As such, evaluation procedures should utilize appropriate hybrid evaluation functions to determine the fitness of candidate solutions. The functions should be able to measure the desired quality of candidate solutions, while considering the fuzzy and sometimes conflicting goals. In the presence of multiple criteria, with a solution space R, each objective function $f_i(s)$ $(i = 1, \ldots, n)$, $s \in R$, is mapped into a corresponding normalized fuzzy function $\mu_i(s)$ based on a suitable membership function. The resulting fitness function, F_t, at iteration t should also be a normalized function of its n constituent normalized functions denoted by $\mu_i(i = 1, \ldots, n)$. Therefore, F_t can be evaluated using a fuzzy multifactor evaluation method as follows:

$$F_t = \sum_{i=1}^{n} w_i \mu_i(s) \tag{4.1}$$

where n is the number of functions $\mu_i(s)$ $(i = 1, \ldots, n)$; w_i is the weight of μ_i; s is a candidate solution at iteration t.

Besides the above method, the normalized fuzzy function $\mu_i(s)$ can be mapped to a single fitness function using the min operator "$\wedge$," according to the following expression:

$$F_t = \left(\frac{\mu_1(s)}{w_1} \wedge 1\right) \wedge \left(\frac{\mu_2(s)}{w_2} \wedge 1\right) \wedge \ldots \wedge \left(\frac{\mu_n(s)}{w_n} \wedge 1\right) \tag{4.2}$$

where $w_i \in (0, 1]$; n is the number of functions $\mu_i(s)$ $(i = 1, \ldots, n)$; s is a candidate solution at iteration t.

4.3.3.2 Fuzzy Goal-Oriented Fitness Evaluation

In a fuzzy multi-criteria environment, the main goal is to find a set of satisfactory Pareto-optimal solutions. Conventional solution approaches search for trade-offs between objectives based on a weighted sum function. When using a weighted sum approach, the main challenge is the determination of weights of individual objectives. A goal-oriented search and optimization approach are preferable, where the best solution should satisfy, as much as possible, a vector of desired fuzzy goals.

Suppose that $x \in S$, where S denotes the solution space generated by FGGA. Let $f_{max} = (f_1^{max}, f_2^{max}, \ldots, f_n^{max})$ denote a vector of user-defined upper bounds for the objective functions, so that, for all i, $f_i^{max} \geq f_i(x)$, for all $x \in S$. Similarly, let $f_{min} = (f_1^{min}, f_2^{min}, \ldots, f_n^{min})$ denote a vector of user-defined upper bounds for the objective functions, so that, for all i, $f_i^{max} \geq f_i(x)$, for all $x \in S$. Then, the desired solutions are in the region defined by $S' = f_i(x) \in [f_i^{min}, f_i^{max}]$, for all i. In real-world practice, f_{min} and f_{max} might not be that attainable. However, the solutions defined by the region are satisfactory. For illustration, suppose there are two objective functions f_1 and f_2, then the region of desired solutions appear as shown in Fig. 4.4.

In view of the above, the desired solution region is a fuzzy set which can be described by a fuzzy goal-based fitness function as follows:

Fig. 4.4 Desired solution
space

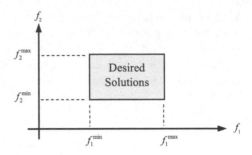

$$\tilde{f} = \begin{cases} \sum_{i=1}^{n} w_i \cdot \mu_i(x), & \text{If } x \in R \\ 0, & \text{If otherwise} \end{cases} \qquad (4.3)$$

where $\sum_i w_i = 1$ and $w_i \geq 0$ and the membership function $\mu_i(x) \in [0,1]$ is defined
as follows:

$$\mu_i(x) = \frac{f_i^{\text{max}} - f_i(x)}{f_i^{\text{max}} - f_i^{\text{min}}}, \qquad \text{For all } i \qquad (4.4)$$

It follows that the higher the values of $\mu_i(x)$, the higher the values of $\tilde{f}_i$, and the
fitter the candidate solution. Expert users can define their choices or preferences in
the upper and lower bound vectors f_{min} and f_{max}, respectively. These can be adjusted
to influence the search process according to expert opinion.

In addition to fuzzy goal-oriented fitness evaluation, fuzzy adaptive genetic
operators have to be introduced in order to effectively explore and exploit the
desired solution space.

4.3.4   Fuzzy Genetic Operators

It has been noted that the behavior of the grouping genetic algorithms (GGA) is
heavily influenced by the trade-off between exploration and exploitation during the
global search process. The magnitude of genetic parameter values plays a very
important role on the iterative search process and, thus, influences the quality of the
final computational results.

4.3.4.1   Fuzzy Controlled Genetic Parameters

The control of the values of genetic parameters, such as convergence measure,
diversity measure, crossover probability, mutation probability, and inversion
probability, is highly influential to the performance of GGA. Their values are an

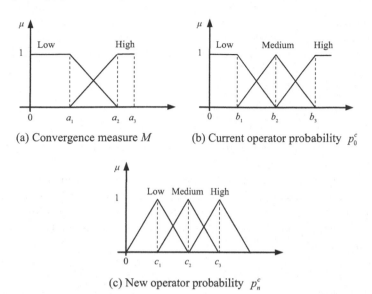

(a) Convergence measure M (b) Current operator probability p_0^c

(c) New operator probability p_n^c

Fig. 4.5 Fuzzy membership functions for input and output of operator probability

essential input to the fuzzy logic control process. The proposed fuzzy logic control uses two inputs, namely:

1. The current convergence measure $M_t \in [0, 1]$ or the current divergence measure $D_t \in [0, 1]$, usually set in [0, 1], and mapped into linguistic values $D_t = \{$Low, High$\}$; and
2. The current value of a genetic operator parameter p_0, for instance, crossover probability, which is mapped into specific linguistic value $p_0 = \{$Low, Medium, High$\}$.

The output is the new genetic operator value p_n, associated with specific linguistic values $p_n = \{$Low, Medium, High$\}$, and is used by the genetic operator in the next generation. The general fuzzy control relationships and the associated membership functions are illustrated in Fig. 4.5. The intervals for the input [a_1, a_2, a_3] and [b_1, b_2, b_3], and output [c_1, c_2, c_3] are derived from the expert opinion, for each genetic parameter. Specific fuzzy rules are generated for specific parameters.

The descriptions of the emerging fuzzy controlled genetic parameters are discussed further.

Convergence Measure

Convergence, as opposed to divergence, is the tendency for characteristics of population chromosomes to stabilize to a common state or solution. In other words, convergence is the tendency for the optimization search process to approach a particular solution. For a satisfactory solution, premature convergence should be avoided. Therefore, a measure of convergence should be defined. Assuming a minimization problem, a convergence measure M is defined by the following expression:

$$M_t = \frac{f_b^c}{f_b^p} \qquad (4.5)$$

where f_b^c is the fitness of the current best solution and f_b^p is the fitness of the best solution in the last τ generations.

It follows that if the value of M_t is high, then convergence is high and little or no improvement was made during the last τ generations. Conversely, if M_t is low, then the algorithm found a much better solution. Since M_t is rather imprecise in practice, fuzzy values, such as low, medium, and high, can be assigned to the parameter for fuzzy control.

Diversity Measure

Population diversity is a phenotype property which measures the level of closeness of population of chromosomes. Like the convergence measure M_t, diversity measure can be formulated and used to control the search and optimization process of the algorithm. A diversity measure D_t can be defined as follows:

$$D_t = \frac{f_t^{max} - f_t^{ave}}{f_t^{max}} \qquad (4.6)$$

where f_t^{max} is the maximum fitness value at generation t and f_t^{ave} is the average fitness value at generation t.

In practice, the numerical values of the diversity measure are mapped to a domain, such as low, medium, and high, in order to control the self-adaptive genetic parameters, based on a defined fuzzy membership function. If D_t is low, this shows that the convergence has taken place, and if D_t is high, then there is a high level of population diversity.

Crossover Probability

The crossover probability (p^c) controls the likelihood of crossover of a candidate solution. This parameter is the most important probability among other genetic probabilities such as mutation and inversion probabilities. The higher the crossover probability p^c, the higher the exploration process, and the higher the number of new solutions generated. However, as p^c increases, solutions can be disrupted too early with little exploitation.

To control the search and optimization process, the current crossover probability p_0^c can be used as input to fuzzy control of the crossover process. A new appropriate crossover probability p_n^c can be generated for the next generation.

Mutation Probability

Mutation probability p^m influences the rate of mutation of individual candidate solutions; the higher the probability, the higher the exploitation rate of the solution. However, with increased mutation probability, the algorithm nearly resembles random search which may lead to loss of the desired balance between random search and guided search. It is therefore important to regulate the values of mutation probability in order to facilitate the global optimization process.

During the search and optimization process, the current mutation value p_0^m can be used as input for creating a new appropriate mutation probability p_n^m through fuzzy control. Fuzzy membership functions and fuzzy if-then rules can be formulated in terms of previous mutation probability values.

Inversion Probability

Inversion is a critical GGA operator whose purpose is to probabilistically overturn some of the candidate chromosomes in order to enhance the crossover operation. The control of the inversion probability p^i is critical to the search and optimization process. However, in practice, the control process is difficult and imprecise, so much that the use of expert knowledge is the most viable option. By applying fuzzy control, the current inversion probability p_0^i is used as input for creating a new and appropriate probability p_n^i. Fuzzy membership functions and fuzzy if-then rules are created using expert opinion.

4.3.4.2 Fuzzy Logic Controlled Crossover

In intervals of τ generations, the FLC is triggered to compute a new value of crossover probability, considering its value during the last τ generations and the improvement achieved on the fitness of the current best solution found f_b.

Inputs and Outputs

For fuzzy crossover control, there are two inputs to the fuzzy logic control model, namely (i) the current crossover probability, p_0^c, whose linguistic values are {Low;

Medium; High}, and (ii) the convergence measure M with a set of linguistic values {Low; High}. The output is the new p_n^c, which should be used in the next τ generations. The set of linguistic labels associated with the output is {Low; Medium; High}. As was demonstrated in Sect. 4.3.4.1, the intervals for the input $[a_1, a_2, a_3]$ and $[b_1, b_2, b_3]$, and output $[c_1, c_2, c_3]$ are derived from expert opinion.

Fuzzy Rule Base

Table 4.1 shows the fuzzy rule base for the proposed FLC. Two general heuristics associated with the rule base are as follows: (i) if there is any improvement in f_b, then decrease p^c, and (ii) if no improvement, then increase p^c. The first holds for rules 1 and 2, and the second holds for rules 4–6. This ensures that p^c values that are too low (promoting premature convergence) or too high (not allowing convergence to obtain better solutions) are avoided.

4.3.4.3 Fuzzy Logic Controlled Mutation

As in the crossover operator, the FLC is triggered every τ generations at which the mutation value p^m is adjusted according to its value in the last τ generations and the improvement f_b in fitness of the current best solution.

Inputs and Outputs

The inputs to fuzzy mutation control are as follows:(i) the current mutation probability p_0^c, with linguistic values set as {Low; Medium; High}, and (ii) the convergence measure M with linguistic values {Low; High}. The output p_n^m, associated with linguistic values {Low; Medium; High}, is used for the next τ generations. The intervals for the input $[a_1, a_2, a_3]$ and $[b_1, b_2, b_3]$, and output $[c_1, c_2, c_3]$ are derived from expert opinion, as indicated in Sect. 4.3.4.1.

Table 4.1 Fuzzy rule base for control of crossover probability p^c

Rule	M_t	p_0^c	p_n^c
1	High	Low	Medium
2	High	Medium	High
3	High	High	Low
4	Low	Low	Low
5	Low	Medium	Low
6	Low	High	Medium

Fuzzy Rule Base

Table 4.2 presents the fuzzy rule base for fuzzy mutation control. The rules are derived from two heuristics: (i) If there is any improvement in f_b, then decrease p^m, and (ii) if there is no improvement, then increase p^m.

The fuzzy rule base is supposed to ensure that p^m values that are too low (which promotes premature convergence) or too high (which hinders convergence to toward better solutions) are avoided.

4.3.4.4 Fuzzy Logic Controlled Inversion

As in other genetic operators, a new inversion probability is computed every τ generations, based on its past value in the last τ generations and the improvement achieved on best solution f_b.

Inputs and Outputs

Two inputs are used for fuzzy logic control: (i) the current inversion probability, p_0^i, with linguistic values {Low; Medium; High}, and (ii) the divergence measure M with linguistic values {Low; High}. The output, p_n^i, used for the next τ generations, is associated with linguistic labels {Low; Medium; High}. The intervals for membership functions of the input $[a_1, a_2, a_3]$ and $[b_1, b_2, b_3]$, and output $[c_1, c_2, c_3]$ are derived from expert opinion.

Fuzzy Rule Base

A fuzzy rule base is developed for inversion control, based on two heuristic facts, namely: (i) Whenever f_b improves, the value of p^i should be reduced, and (ii) when there is no improvement in f_b, the value of p^i should be increased to promote diversity and crossover operations in the next generations. The first heuristic applies to fuzzy rules 1 and 2, and the second applies to fuzzy rules 4–6 as summarized in Table 4.3.

Table 4.2 Fuzzy rule base for control of mutation probability p^m

Rule	M_t	p_0^m	p_n^m
1	High	Low	Medium
2	High	Medium	High
3	High	High	Low
4	Low	Low	Low
5	Low	Medium	Low
6	Low	High	Medium

Table 4.3 Fuzzy rule base for control of inversion probability p^i

Rule	D_t	p_0^i	p_n^i
1	Low	Low	Medium
2	Low	Medium	High
3	Low	High	Low
4	High	Low	Low
5	High	Medium	Low
6	High	High	Medium

The above setting ensures that premature convergence (due to very low p^i values) and excessive divergence (due to very high p^i values) are always under control during the search and optimization process.

The next section deals with dynamic fuzzy adaptive operators which adjust themselves based on an evaluation of online input variables.

4.3.5 Fuzzy Dynamic Adaptive Operators

Fuzzy dynamic adaptive operators utilize fuzzy theory concepts to adapt themselves to the state of the population solutions, taking into account the iteration time t. Thus, the dynamic self-adaptive features of the operators rest on the premise that the respective probabilities of crossover, mutation, and inversion, that is p^c, p^m, and p^i, must be adjusted online according to the fitness levels of the chromosomes in the population, as well as the time. The following input variables are instrumental in this regard:

f_{ave} the average fitness in the current population;
f_{max} the maximum fitness of the current generation;
f_{min} the minimum fitness of the current population; and
t the time, or generation count, out of the maximum count T.

The above-listed input variables can provide the genetic operators an ability to adjust themselves on the run. Based on the variable f_{ave}, the population is subdivided into two subpopulations: (i) the weak and (ii) the fittest, whereupon suitable genetic operators may be applied. For a chromosome with fitness f, the genetic parameters of each operator are adjusted according to its relative quality and the time t. The relative quality Q is defined as follows:

$$Q = \frac{f_{max} - f}{f_{max} - f_{ave}} \tag{4.7}$$

The above expression suggests the relative quality of the chromosome in relation to the population. Adaptive crossover, mutation, and inversion operators can be defined based on the premise that the fittest should be preserved into the next generation, while the weak are transmuted.

4.3.5.1 Fuzzy Dynamic Adaptive Crossover

By classifying chromosomes into two subpopulations, special crossover operators can be assigned to each subpopulation, depending on the features of the specific crossover operators and the characteristics of the subpopulation. A rule base can then be constructed in this regard, for instance,

If $\{f < f_{ave}\}$ **Then** apply two-point crossover

If $\{f > f_{ave}\}$ **Then** apply single-point crossover.

Here, the single-point crossover swaps two individual groups from two different chromosomes at a dynamic probability p^c. Based on the relative quality Q, the general heuristic for the control of the crossover probability p^c is stated thus: (i) If Q is high, then increase p^c, so as to maximize the likelihood of crossing the fittest parent chromosomes to produce even fitter offspring; and (ii) if the value is low, then decrease p^c to avoid the likely weak offspring from the weak parent chromosomes. Therefore, the p^c values are adjusted according to the following rules:

$$p^c = \begin{cases} Q \cdot k_1 e^{-t/T} & \text{If } f \geq f_{ave} \\ k_2 & \text{If } f < f_{ave} \end{cases} \tag{4.8}$$

where $k_1 = 1$ and $k_2 = 0.5$ define the characteristic parameters for the control of the crossover operator for the fittest and weak subpopulations, respectively. Here, the term $e^{-t/T}$ allows the value of p^c to decrease dynamically.

4.3.5.2 Fuzzy Dynamic Adaptive Mutation

Specific mutation operators are applied to each chromosome in the two subpopulations, depending on the features of the mutation operators and the subpopulation. In general, for a chromosome with fitness f,

If $\{\{f < f_{ave}\}$ **Then** apply split and merge mutation

If $\{f \geq f_{ave}\}$ **Then** apply swap mutation

Here, the mutation probability p^m is controlled using the relative quality Q, based on a general heuristic stated thus: (i) If Q is high, then decrease p^m, so as to reduce the likelihood of mutating the fittest chromosomes, but rather preserve them; and (ii) if Q is low, then increase p^m to promote the likelihood of mutating weak chromosomes in order to possibly improve them. As such, p^m values are adjusted according to the following rules:

$$p^m = \begin{cases} Q \cdot k_3 e^{-t/T} & \text{If } f \geq f_{\text{ave}} \\ k_4 & \text{If } f < f_{\text{ave}} \end{cases} \tag{4.9}$$

where $k_3 = 0.7$ and $k_4 = 0.4$ define the characteristic parameters for the controlling mutation of the fittest and weak subpopulations, respectively. Here, the term $e^{-t/T}$ allows the value of p^m to decrease dynamically.

4.3.5.3 Fuzzy Dynamic Adaptive Inversion

The inversion operator overturns genes of chromosomes at a low probability p^i, in order to promote diversity. Specific inversion operators are applied to specific subpopulations, according to the features of each subpopulation and the inversion operators. As a general heuristic rule,

If $\{f < f_{\text{ave}}\}$ **Then** apply the two-point inversion operator;

If $\{f \geq f_{\text{ave}}\}$ **Then** apply the full inversion operator.

Here, the two-point inversion overturns two selected genes of a chromosome, while the full inversion overturns the whole chromosome. Using the relative quality Q, a general heuristic is applied: (i) If Q is high, then increase p^i, in order to promote crossover of the fittest chromosome in the next generation; and (ii) if Q is low, then decrease p^i to discourage crossover of weak chromosomes. As such, the p^i values are adjusted as follows:

$$p^i = \begin{cases} Q \cdot k_5 e^{-t/T} & \text{If } f \geq f_{\text{ave}} \\ k_6 & \text{If } f < f_{\text{ave}} \end{cases} \tag{4.10}$$

where $k_5 = 0.7 = 1$ and $k_6 = 0.5$ define the characteristic parameters for controlling inversion of the fittest and weak subpopulations, respectively. The term $e^{-t/T}$ enforces dynamic decay of p^i over time t.

4.3.6 Termination

The FGGA search and optimization continue iteratively until a termination condition is satisfied: when (1) prespecified maximum number of iterations, T, is reached, or (2) when the current best solution has not improved in a prespecified number of iterations, or (3) when the two conditions 1 and 2 are satisfied simultaneously. The next section presents the potential application areas for the proposed algorithm.

4.4 Potential Application Areas

Problem characteristics that necessitate the application of the FGGA approach in real-world industry include (i) problem complexity due to the curse of dimensionality, common in hard combinatorial problems; (ii) the presence of imprecise fuzzy management aspirations and goals, and other preferences; (iii) the presence of multiple optimization criteria; and (iv) the presence of multiple confliction constraints, or (v) a combination of one or more these characteristics. Problems with imprecise features or qualitative variables are difficult to model using conventional methods. In addition, problems with computational complexities can be addressed more efficiently and effectively by applying fuzzy dynamic adaptive control on genetic operators of the algorithm.

Table 4.4 lists some of the potential areas of application and brief descriptions of their respective complex characteristics. It can be seen that the FGGA procedure is wide applicable to a number of industrial grouping problems from various disciplines, including manufacturing systems, design and human factors engineering, maintenance management, warehouse and distribution systems, logistics management, healthcare systems, education, research and development (R&D), and business and economics.

It is hoped the proposed FGGA can address the complex challenges found in these problems. The next section summarizes the chapter.

Table 4.4 Potential application areas and their characteristics

No.	Grouping problems	Brief description of complexities	Selected references
1	Assembly line balancing	Complex large-scale instances need efficient algorithm	Sabuncuoglu et al. (2000)
2	Job shop scheduling	A hard and highly combinatorial problem; computationally expensive	Chen et al. (2012) and Phanden et al. (2012)
3	Cell formation	A hard highly combinatorial problem; complex constraints	Onwubolu and Mutingi (2001)
4	Container loading	Several complex constraints; multiple objectives; complex loading patterns	Althaus et al. (2007) and Joung and Noh (2014)
5	Heterogeneous fixed fleet Vehicle routing	Imprecise information at planning stage; highly combinatorial	Tutuncu et al. (2010)
6	Fleet size and mix vehicle routing	Complex and highly combinatorial; complex constraints	Liu et al. (2009) and Brandao (2008)
7	Group maintenance planning	Highly combinatorial; computationally expensive, needing efficient algorithms; imprecise cost functions	Do Van et al. (2013) and Gunn and Diallo (2015)

(continued)

Table 4.4 (continued)

No.	Grouping problems	Brief description of complexities	Selected references
8	Handicapped person Transportation	Uncertain travel times; time window preferences	Rekiek et al. (2006)
9	Task assignment	Uncertain task times; worker preferences	Mutingi and Mbohwa (2014a, b, c)
10	Home healthcare scheduling	Imprecise management goals; patient preferences; worker preferences	Mutingi and Mbohwa (2014a, b, c)
11	Multiple traveling salesperson	Highly combinatorial, and computationally expensive, with uncertain travel times.	Kivelevitch and Cohen (2013) and Bektas (2006)
12	Modular product design	Imprecise life cycle cost functions; fuzzy management goals	Yu et al. (2011) and Chen and Martinez (2012)
13	Order batching	Highly combinatorial; computationally expensive	Henn and Wascher (2012) and Henn (2012)
14	Pickup and delivery	Highly combinatorial; computationally expensive; imprecise time windows	Chen (2013)
15	Learners grouping	Ill-defined qualitative learner preferences	Chen et al. (2012) and Baker and Benn (2001)
16	Team formation	Imprecise member choices; management goals;highly combinatorial and computationally expensive	Wi et al. (2009) and Strnad and Guid (2010)
17	Reviewer group construction	Skills levels are qualitative; reviewer choices are vague; management goals are imprecise	Chen et al. (2011)
18	Estimating discretionary accruals	Imprecise economic variables; management estimations largely qualitative	Höglund (2013) and Bartov et al. (2000)

4.5 Summary

Grouping genetic algorithm (GGA) is an effective and efficient algorithm that has been used to solve a number of grouping or clustering problems. The main challenge, however, is the development of the computational procedure for recent complex problems characterized with uncertain and imprecise variables, imprecise goals and preferences, the curse of dimensionality, and other complexities. In practice, it is difficult to understand how genetic parameters interact and how their interactions influence the performance of the algorithm. This is because the interactions are complex and difficult to model in a precise way. Thus, fine-tuning, control, and adaptation of the behavior of genetic parameters, such as divergence, crossover, mutation, and inversion probabilities, are a cause for concern in the research community.

This chapter focused on proposing advances and innovations in the use of fuzzy logic control and dynamic adaptive control, and their incorporation into the GGA mechanism. In FGGA, grouping genetic operators are enriched with fuzzy logic control and other fuzzy theoretic concepts so as to enable the algorithm to accommodate expert choice, opinion, and guidance during its search and optimization processes. The algorithm is expected to be able to handle complex real-world grouping problems with fuzzy characteristics. It is hoped that the fuzzy GGA (FGGA) proposed in this chapter will be effective and efficient for solving real-world grouping problems, even in a fuzzy environment. Prospects for possible application areas in this direction were presented.

References

Arnone S, Dell'Orto M, Tettamanzi A (1994) Toward a fuzzy government of genetic populations. In: Proceedings of the 6th IEEE conference on tools with artificial intelligence, IEEE Computer Society Press, Los Alamitos, CA, pp 585–591

Althaus E, Baumann T, Schömer E, Werth K (2007) Trunk packing revisited. LNCS 4525: 420–430

Bartov E, Gul FA, Tsui JSL (2000) Discretionary-accruals models and audit qualifications. J Account Econ 30(3):421–452

Baker BM, Benn C (2001) Assigning Pupils to Tutor Groups in a Comprehensive School. J Oper Res Soc 52(6):623–629

Bektas T (2006) The multiple traveling salesman problem: an overview of formulations and solution procedures. Omega 34(3):209–219

Brandao J (2008) A deterministic tabu search algorithm for the fleet size and mix vehicle routing problem. Eur J Oper Res 195(3):716–728

Chao XL, Zheng Z, Fan N, Wang XF (1999) A modified genetic algorithm by integrating neural network technology. Pattern Recog Artif Intell 12:486–492

Chen Y, Fan Z-P, Ma J, Zeng S (2011) A hybrid grouping genetic algorithm for the reviewer group construction problem. Expert Syst Appl 38:2401–2411

Chen AL, Martinez DH (2012) A heuristic method based on genetic algorithm for the baseline-product design. Expert Syst Appl 39(5):5829–5837

Chen JC, Wu C-C, Chen C-W, Chen K-H (2012) Flexible job shop scheduling with parallel machines using Genetic Algorithm and Grouping Genetic Algorithm. Expert Syst Appl 39(2012):10016–10021

Chen Y-Y (2013) Fuzzy flexible delivery and pickup problem with time windows. Information Technology and Quantitative Management (ITQM2013). Procedia Comput Sci 17:379–386

Cordon O, Herrera F, Hoffmann F, Magdalena L (2001) Genetic fuzzy systems. Evolutionary tuning and learning of fuzzy knowledge bases. World Scientific

Do Van P, Barros A, Bérenguer C, Bouvard K, Brissaud F (2013) Dynamic grouping maintenance with time limited opportunities. Reliab Eng Syst Safe 120:51–59

Driankow D, Hellendoorn H, Reinfrank M (1993) An introduction to fuzzy control. Springer-Verlag, Berlin

Falkenauer E (1992) The grouping genetic algorithms—widening the scope of the GAs. Belg J Oper Res Stat Comput Sci 33:79–102

Falkenauer E (1994) A new representation and operators for genetic algorithms applied to grouping problems. Evol Comput 2:123–144

Falkenauer E (1996) A hybrid grouping genetic algorithm for bin packing. J Heuristics 2:5–30

Gunn EA, Diallo C (2015) Optimal opportunistic indirect grouping of preventive replacements in multicomponent systems. Comput Ind Eng 90:281–291

Henn S (2012) Algorithms for on-line order batching in an order picking warehouse. Comput Oper Res 39:2549–2563

Henn S, Wäscher G (2012) Tabu search heuristics for the order batching problem in manual order picking systems. Eur J Oper Res 222:484–494

Herrera F, Lozano M (2003) Fuzzy adaptive genetic algorithms: design, taxonomy, and future directions. Soft Comput 7:545–562

Höglund H (2013) Estimating discretionary accruals using a grouping genetic algorithm. Expert Syst Appl 40:2366–2372

Hu, XB, Wu, SF (2007) Self-adaptive genetic algorithm based on fuzzy mechanism. Paper presented at the 2007 IEEE congress on evolutionary computation (CEC2007), pp 25–28

Hu XB, Wu SF, Jiang J (2004) On-line free-flight path optimization based on improved genetic algorithms. Eng Appl Artif Intell 17:897–907

Hu XB, Paolo ED, Wu SF (2008) A comprehensive fuzz-rule-based self-adaptive genetic algorithm. Int J Intell Comput Cybern 1(1):94–109

Joung Y-K, Noh SD (2014) Intelligent 3D packing using a grouping algorithm for automotive container engineering. J Comput Des Eng 1(2):140–151

Kivelevitch E, Cohen K (2013) Manish Kumar. A market-based solution to the multiple traveling salesmen problem. J Int & Robotic Syst 72(1):21–40

Li J, Kwan RSK (2001) A fuzzy simulated evolution algorithm for the driver scheduling problem. Proceedings of the 2001 IEEE Congress on Evolutionary Computation, IEEE Service Center, pp. 1115–1122

Liu S, Huang W, Ma H (2009) An effective genetic algorithm for the fleet size and mix vehicle routing problems, Transport Res Part E 45:434–445

Mutingi M, Mbohwa C (2014a) Home health care staff scheduling: effective grouping approaches. IAENG transactions on engineering sciences—special issue of the international multi-conference of engineers and computer scientists, IMECS 2013 and world congress on engineering, WCE 2013, CRC Press, Taylor & Francis Group, pp 215–224

Mutingi M, Mbohwa C (2014b) Multi-objective homecare worker scheduling—a fuzzy simulated evolution algorithm approach. IIE Trans Health Syst Eng 4(4):209–216

Mutingi M, Mbohwa C (2014c) A fuzzy-based particle swarm optimization approach for task assignment in home healthcare. S Afr J Ind Eng 25(3):84–95

Mutingi M, Mbohwa C (2015) Nurse scheduling: a fuzzy multi-criteria simulated metamorphosis approach. Eng Lett 23(3):222–231

Onwubolu GC, Mutingi M (2001) A genetic algorithm approach to cellular manufacturing systems. Comput Ind Eng 39:125–144

Phanden RK, Jain A,Verma R (2012) A genetic algorithm-based approach for job shop scheduling. J Manuf Technol Manage 23(7):937–946

Rekiek B, Delchambre A, Saleh HA (2006) Handicapped Person Transportation: An application of the Grouping Genetic Algorithm. Eng Appl Artif Intell 19(5):511–520

Sabuncuoglu I, Erel E, Tanyer M (2000) Assembly line balancing using genetic algorithms. J Intell Manuf 11(3):295–310

Strnad D, Guid N (2010) A Fuzzy-Genetic decision support system for project team formation. Appl Soft Comput 10(4):1178–1187

Tutuncu GY (2010) An interactive GRAMPS algorithm for the heterogeneous fixed fleet vehicle routing problem with and without backhauls. Eur J Oper Res 201(2):593–600

Wi H, Oh S, Mun J, Jung M (2009) A team formation model based on knowledge and collaboration. Expert Syst Appl 36(5):9121–34

Yu S, Yang Q, Tao J, Tian X, Yin F (2011) Product modular design incorporating life cycle issues - Group Genetic Algorithm (GGA) based method. J Cleaner Prod 19(9–10):1016–1032

Part III
Research Applications

Chapter 5
Multi-Criterion Team Formation Using Fuzzy Grouping Genetic Algorithm Approach

5.1 Introduction

Efficient utilization of knowledge and skills is essential in every organization, particularly where project management teams, multifunctional workforce teams, task force teams, research and development (R&D) teams, and research-oriented teams are required (Agustin-Blas et al. 2011; Wi et al. 2009a, b; Tavana et al. 2007; Keeling et al. 2007; Wi et al. 2009a, b). In such situations, human resources are organized into teams, according to their skills, so as to perform tasks that require specific skills and knowledge (Fitzpatrick and Askin 2005). The goal is to establish team members from the available pool of human resources in order to maximize, as much as possible, the overall performance of the teams. In other words, the objective is to develop organizational teams or to specialize groups of staff on a given subject resource (Agustin-Blas et al. 2009). In this view, all members in a given group have similar skills of the resource. As such, the problem can be described as team formation without leaders (Agustin-Blas et al. 2011; Tseng et al. 2004).

Recently, an appreciable number of researchers have attempted to solve the problem of team formation without leaders. There are three basic assumptions associated with the team formation: (i) The levels of available skills are precisely known a priori, (ii) the teams take into account all the skills of the team members, and (iii) the utilization of skills in each team should be maximized as much as possible. However, it can be seen that in the presence of high variety skill categories, the problem becomes complex. In practice, the rating of skill levels is imprecise. Moreover, rather than maximizing total knowledge in a team, other criteria, such maximizing total knowledge for each resource, may need to be considered in real-world practice. Extant studies do not seem to address this gap adequately.

In addressing this gap, there is a need to model the fuzzy characteristics, as well as the multiple criteria inherent in the problem. However, addressing these

© Springer International Publishing Switzerland 2017
M. Mutingi and C. Mbohwa, *Grouping Genetic Algorithms*,
Studies in Computational Intelligence 666,
DOI 10.1007/978-3-319-44394-2_5

situations is a complex undertaking, requiring advanced solution approaches such as evolutionary algorithms, metaheuristic algorithms, and fuzzy multi-criterion decision-making approaches. A hybrid approach to the problem, combining grouping genetic algorithms and fuzzy evaluation techniques is suggested.

In view of the above gaps and realizations, the purpose of this research is to develop a fuzzy grouping metaheuristic algorithm for the team formation problem in a fuzzy multi-criterion environment. In this endeavor, the following objectives are pursued as follows:

1. Define and formulate the multi-criterion team formation problem in a fuzzy environment;
2. Develop a fuzzy multi-criterion grouping genetic algorithm approach for the problem; and
3. Carry out computational experiments to test the efficiency and effectiveness of the approach.

The next section presents the related literature. Section 5.3 provides a description of the multi-criterion team formation problem. A fuzzy grouping genetic algorithm is presented in Sect. 5.4. Section 5.5 provides illustrative computation tests, results, and discussions. Finally, Sect. 5.6 presents a summary of this chapter.

5.2 Related Approaches

In recent years, a few researchers have considered team formation problems with imprecise or fuzzy characteristics. It has been noted that, oftentimes, team member preferences, management aspirations, decision maker's intuition, and other constraints are often fuzzy. Tseng et al. (2004) proposed a novel approach to multi-functional project-based team formation, based on fuzzy set theory. The main aim was to match customer's requirements and engineers skills or knowledge. The approach incorporated on fuzzy set theory and gray decision theory to deal with ambiguities associated with multi-functional team formation where the relationship boundary between customer requirements and project characteristics is not crisp.

Strnad and Guid (2010) developed a fuzzy set theoretic genetic decision support system to solve a project team formation problem, taking into account different skills in different tasks. The model seeks to optimize selection of multiple project teams with conflicting requirements using a specialized island genetic algorithm with a mixed crossover operator. Multiple conflicting criteria are converted into a single objective function which is should be maximized. Dereli et al. (2007) presented a case study concerned with the formation of quality audit team, based on auditor skills, to audit various organizations with varying audit requirements, located at different places. Each audit team must fulfill the skills set required for a specific audit. The authors proposed a fuzzy mathematical program for the problem and solved it using a simulated annealing-based algorithm.

Fitzpatrick and Askin (2005) considered a worker team formation problem in a manufacturing setting with multifunctional skill requirements. The authors considered innate worker tendencies, interpersonal skills, and technical skills. Measures of innate tendencies are used together with experiential results on the desirable team composition aimed at maximizing the effectiveness of the teams. A mathematical programming model was proposed for the team formation problem and solved using a heuristic approach. Recently, Agustin-Blas et al. (2011) presented a group technology-based model for the team formation problem without leaders. The approach is derived from the group technology concepts found in cell manufacturing systems (Onwubolu and Mutingi 2001). The problem typically arises in research and development-oriented organizations and teaching institutions where teaching groups need to be formed. A parallel hybrid grouping genetic algorithm is presented to solve synthetic problem instances and a case example concerned with the formation teaching groups.

Considering the above work, it will be desirable to develop efficient multi-criterion heuristic algorithms which incorporate fuzzy set theoretic concepts to model imprecise human tendencies, preferences, choices, and decision maker's intuition. Furthermore, incorporating fuzzy operators in the algorithm can enhance its search and optimization procedure. With the possible addition of group technology concepts, the resulting algorithm is expected to be more efficient and effective.

5.3 The Multi-Criterion Team Formation Problem

In this section, the team formation problem is modeled from a fuzzy multi-criterion perspective. A description of the problem is presented, leading to fuzzy model development.

5.3.1 Problem Description

The multi-criterion team formation problem (TFP) can be described based on the group technology (GT) concepts. Consider a set of E staff members and a set of R resources, where the term resource is taken in a general sense, for instance, to refer to a type of machine, or a subject area in a college (Augustin-Blass et al. 2011). Each staff member e ($e = 1, \ldots, E$) has specific skills represented by skills matrix K whose integer-valued entries k_{er} ($j = 1, \ldots, R$) reflect the qualitative level of knowledge or skill in a particular resource, where, typically, $k_{er} \in [0, 5]$. The aim is to obtain a diagonal matrix K', representing efficient teams of staff members based on a set of criteria. Figure 5.1 shows an example of a knowledge matrix K with 13 staff members and 10 resources, where part (a) is the initial knowledge matrix K and part (b) is the diagonalized matrix K'.

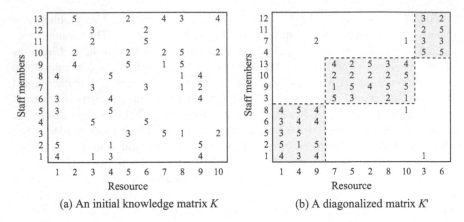

(a) An initial knowledge matrix K (b) A diagonalized matrix K'

Fig. 5.1 An illustration of a team formation matrix diagonalization

Here, the performance measure defined for the binary matrix such as in machine-component matrix in group technology-based cell formation cannot apply. Thus, instead, a different set of performance measures are proposed for this application.

5.3.2 Fuzzy Multi-Criterion Modeling

Fuzzy multiple criteria are often modeled using membership functions, such as generalized bell, interval-valued, gaussian, triangular and trapezoidal membership functions (Sakawa 1993a, b; Mutingi and Mbohwa 2016). Linear membership functions have been known to provide good quality solutions with much ease (Sakawa 1993a, b; Chen 2001; Delgado et al. 1993). Figure 5.2 shows a symmetrical fuzzy triangular membership function. By this membership function, the satisfaction level is represented by a fuzzy number $A \langle c, a \rangle$, where c denotes the center of the fuzzy parameter with width a.

Based on the triangular fuzzy membership representation, the corresponding membership of x in the fuzzy set A, $\mu_A : X \rightarrow [0, 1]$, is given by the following expression:

Fig. 5.2 Triangular membership function triangular membership function

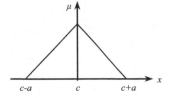

$$\mu_A(x) = \begin{cases} 1 - \frac{|c-x|}{a} & \text{if } c - a \le x \le c + a \\ 0 & \text{if otherwise} \end{cases} \qquad (5.1)$$

Depending on the problem setting, other membership functions specific to the situation can be formulated.

5.3.2.1   Fuzzy Model

To improve the quality of the candidate solutions, the mean knowledge that the team members have about the resources assigned to the team must be as satisfactory as possible. The mean knowledge of the team, f_g, can be estimated by the following expression:

$$f_g \cong \left(\sum_{r=1}^{R} \sum_{e=1}^{E} x_{rg} y_{eg} \tilde{k}_{er} \right) \bigg/ \left(\sum_{r=1}^{R} \sum_{e=1}^{E} x_{rg} y_{eg} \right) \qquad \forall g \qquad (5.2)$$

where $X_{rg} = 1$ if resource r is assigned to team g, and otherwise, 0; $Y_{eg} = 1$ if staff member e is assigned to team g, and otherwise 0.

In addition to the above criterion, it is essential to ensure that the mean knowledge f_e of each staff e in that particular team is as satisfactory as possible. Therefore, f_e can be estimated as follows:

$$f_e \cong \left(\sum_{g=1}^{G} \sum_{r=1}^{R} x_{rg} y_{eg} \tilde{k}_{er} \right) \bigg/ \left(\sum_{g=1}^{G} \sum_{r=1}^{R} x_{rg} y_{eg} \right) \qquad \forall e \qquad (5.3)$$

This expression encourages assignment of individual staff to groups in which they will exercise the best of their skills. As a result, the staff will be more productive. To further improve the quality of the solutions, the mean knowledge f_r of the team about a specific resource r should be as satisfactory as possible. Thus, the mean knowledge f_r can be estimated by the expression:

$$f_r \cong \left(\sum_{g=1}^{G} \sum_{e=1}^{E} x_{rg} y_{eg} \tilde{k}_{er} \right) \bigg/ \left(\sum_{g=1}^{G} \sum_{e=1}^{E} x_{rg} y_{eg} \right) \qquad \forall r \qquad (5.4)$$

Fuzzy set theory permits gradual assessment of membership, defined in terms of a suitable membership function that maps to the unit interval [0, 1]. A number of membership functions such as generalized bell, gaussian, triangular, and trapezoidal can be used to represent the fuzzy membership. Though various functions can be used, it has been shown that linear membership functions can provide equally good quality solutions with much ease Sakawa (Sakawa 1993a, b). The triangular and trapezoidal membership functions are widely acceptable (Chen 2001; Delgado et al. 1993).

Therefore, in this chapter, we use triangular membership functions to define the fuzzy objectives.

5.4   A Fuzzy Grouping Genetic Algorithm Approach

Fuzzy grouping genetic algorithm (FGGA) is an extension of grouping genetic algorithm (GGA) in which at least one of its genetic operators (selection, crossover, mutation, and inversion) use fuzzy set theoretic techniques. In this application, fuzzy evaluation and selection are used to evaluate the fitness of solutions, from a multiple criteria perspective. Crossover, mutation, and inversion also utilize fuzzy set theoretic concepts.

5.4.1   Group Encoding Scheme

In the team formation problem, we assume the fixed-length encoding scheme, since, in most real-world practices, the desired number of teams is often known a priori. Thus, assume that the items represent the staff members, $(i = 1, 2, …, m)$, who are to be clustered into g groups (teams) $(j = 1,2, …, g)$. Every staff member can belong to any group, subject to group size restrictions. To illustrate further, consider a situation where 8 members are to be allocated to 3 groups, where each group should have at least 2 members. Figure 5.3 shows a group encoding for a typical solution to this situation. Here, group identified by 1, 2, and 3, which signifies chromosome $[1 \quad 2 \quad 3]$, and correspond to team members $\{1, 7\}$, $\{4, 3, 8\}$, and $\{2, 5, 6\}$, respectively.

Notably, each item (member) can only be allocated to one group. The representation is order-independent, implying that the sequence of items in each group is not important. The grouping operators take advantage of the group structure of the problem, ensuring that the group spirit of the candidate solutions is preserved. This has a positive influence on the search and optimization processes (Kashan et al. 2015; Falkenauer 1998).

5.4.2   Initialization

The initial population is generated randomly, that is, the heuristic randomly places items into each of the g groups, subject to group size limit. In turn, the resources are

Fig. 5.3 Group encoding scheme for the team formation problem

Members:	1,7	4,3,8	2,5,6
Group:	1	2	3

assigned heuristically, where each resource is allocated to a group in which most of its skills are found. For instance, from Fig. 5.1b, it is shown that resource 9 has 4 skills in group 1, which is far more than just 1 skill in group 3. It follows that resource 9 should be in group 1, and so on.

Algorithm 1 Initialization procedure

1. **Input**: $p, maxgen, p^c, p^m, p^i, n, g$;
2. Initialize, $n = 0$;
3. **Repeat**
4.   Select item $i*$ at random;
5.   Select group $j*$;
6.   **If** (group size < maximum) **Then**
7.   assign item $i*$ to group $j*$ to an of g groups;
8. $n = n + 1$;
9.   **End**
10. **Until** $(n = m)$
11. **For** $i = 1$ to m **do**   //assign resource i to group j with most of its skills
12.   Resource group[i] = argmax[resource$_{ij}$];
13. **End**
14. **End**
15. **Return:** Population P

The population chromosomes are then evaluated using a fuzzy multi-criterion method.

5.4.3   Fuzzy Evaluation

Fuzzy evaluation of population chromosomes is conducted based on three criteria, that is, (i) mean knowledge of the team f_g, (ii) mean knowledge of each staff f_e in a team, and (iii) mean knowledge f_r about a specific resources, as formulated in the previous section. The fuzzy membership function in (5.1) is used for the evaluation of chromosomes. Thus, for the mean knowledge of the team f_g, assume that the limits in the membership function is as follows:

$$\mu_1 = \mu_A(f_g) \tag{5.5}$$

where c and a are the desired mean and the acceptable deviation of the f_g from the mean. The parameters are normally derived from the expert opinion.

For the mean knowledge of each staff f_e in a team, let the fuzzy membership function be $\mu_e = \mu_A(f_e)$. This implies that for all the employees, $e = 1, ..., m$, the membership function can be determined by the expression:

$$\mu_2 = \sum_{e=1}^{m} \mu_e \tag{5.6}$$

In a similar manner, the mean knowledge f_r of the team about a specific resource r is $\mu_r = \mu_A(f_r)$. Then, the expression for the membership function across all the resources $r = 1$ to R, is as follows:

$$\mu_3 = \sum_{r=1}^{R} \mu_r \tag{5.7}$$

The three membership functions μ_1, μ_2, and μ_3 can be combined into a single normalized function. To further incorporate the decision maker intuition and expert knowledge, and model flexibility, a set of user-defined weights $w = \{w_1, w_2, w_3\}$ are introduced. The resulting single objective function is given by the following expression:

$$\mu = \sum_{z=1}^{3} w_z \mu_z \tag{5.8}$$

where $w_z \in [0, 1]$ denotes the user-defined weight of the tth membership function, such that $\sum w_z = 1$; μ_z denotes the membership function or the degree of satisfaction of the zth objective.

5.4.4 Selection and Crossover

The algorithm selects the best performing chromosomes into a mating pool or temporal population, called *tempp*. Among the well-known selection mechanisms, the remainder stochastic sampling without replacement is the most widely accepted (Goldberg 1989; Mutingi and Mbohwa 2016). In this mechanism, each chromosome k is selected and stored in *tempp* according to its expected count e_k given by the expression:

$$e_k = a \cdot \mu(k) \Big/ \sum_{k=1}^{p} \left(\frac{\mu(k)}{p} \right) \tag{5.9}$$

where $\mu(k)$ is the score function of the k_{th} chromosome, $a \in [0,1]$ is an adjustment parameter, and p is the population size. Under this strategy, each chromosome obtains copies equal to the integer part of e_k, plus additional copies at a probability equivalent to the fractional part of e_k. The selection mechanism ensures that the best performing candidates are selected with higher probability for the crossover operation.

A two-point crossover strategy is adopted from the conventional two-point crossover mechanism in the basic genetic algorithm (GA). The crossover strategy is summarized into four stages as outlined in the Algorithm 2:

Algorithm 2 Two-point crossover procedure

Stage 1 Randomly selected two parent chromosomes, P_1 and P_2, and select crossing sections for the two chromosomes

Stage 2 Cross P_1 and P_2 by interchanging the crossing sections of the parent chromosomes. Obtain two offspring O_1 and O_2, which may have repeated items (called doubles) or missing items (called misses)

Stage 3 Eliminate the doubles, using a constructive repair mechanism , avoiding the crossed items

Stage 4 Insert the missing items into groups, beginning from where doubles were eliminated, subject to group size limits and other specific side constraints

The process is repeated until a population of new offspring (called *spool*), of size $s = p^c \times p$, is created. To illustrate this algorithm, consider two parent chromosomes (i) $P_1 = [\,1 \quad 2 \quad 3 \quad 4\,]$, corresponding to groups $\{1, 7\}$, $\{3, 5\}$, $\{2, 6\}$, and $\{8, 4, 9\}$, and (ii) $P_2 = [\,5 \quad 6 \quad 7 \quad 8\,]$, corresponding to groups of items $\{8, 7, 5\}$, $\{3, 2\}$, $\{6, 9\}$, and $\{1, 4\}$, as shown in Fig. 5.4.

After crossover, two offspring O_1 and O_2 are produced. After the completion of the crossover operation, a new population, *newpop*, is formed by elitist replacement.

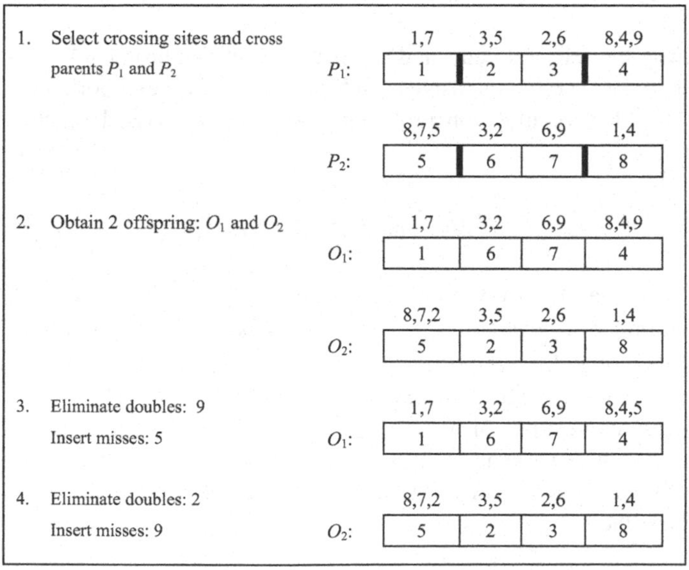

Fig. 5.4 Group crossover operation

This implies that the best performing chromosomes in the current generation t is automatically passed on to the next generation $t + 1$. The rest of the population is built up by from the best performing chromosomes from *spool* and the old population *oldpop*. By so doing, the best performing solutions are always preserved. The *newpop* is then subjected to mutation.

5.4.5 Mutation

Mutation encourages slight probabilistic perturbations on the population chromosomes, exploiting the neighborhoods of candidate solutions to improve their fitness. Deriving from the swap mutation from the basic GA, the GGA swap mutation randomly interchanges selected genes within a chromosome, with a low probability p^m. However, unlike in the basic GA, the swap mutation mechanism operates on the groups rather than on individual items of the chromosome. In case of infeasibilities due to violation of hard constraints, a repair mechanism is utilized. The swap mutation procedure can be summarized into a four-stage algorithm as follows:

Algorithm 3 The swap mutation procedure

Step 1   At a probability p^m, successively select a chromosome from the current population;

Step 2   Randomly choose two different groups from the selected chromosome;

Step 3   Randomly select two items, one from each group of the selected groups; and

Step 4   Swap the selected items and repair the resulting chromosome if necessary.

In order to enhance the intensive search process in the neighborhood of the best solutions, an adaptive mutation probability is computed based on the following expression:

$$p^m(t) = p_0^m + \frac{t}{T}\left(p_f^m - p_0^m\right) \tag{5.10}$$

where $p^m(t)$ is the crossover probability at generation t; T is for the predefined maximum number of generations; and p_0^m and p_f^m are the initial and final values of the mutation probability, respectively. Figure 5.5 further demonstrates the swap mutation mechanism, based on an example of a chromosome $P_1 = \begin{bmatrix} 1 & 7 & 3 & 4 \end{bmatrix}$ with corresponding groups of items $\{1, 3, 7\}$, $\{6, 5\}$, $\{2, 9\}$, and $\{2, 8\}$.

In the first place, two genes, that is, groups 7 and 4, are selected randomly. This is followed by successive random selection of pairs of items from the two groups and swapping them.

1. Select chromosome P1 for mutation, with probability p^m	P_1:	1,3,7	6,5	2,9	2,8
		1	7	3	4

2. Randomly select two groups: 7 and 4	P_1:	1,3,7	6,5	2,9	2,8
		1	7	3	4

3. Randomly select items from the two groups: 5 and 2		1,3,7	6,5	2,9	2,8
		1	7	3	4

4. Swap selected items: 2 and 5 Repair if need be	P_1:	5,3	6,2	2,9	5,8
		1	7	3	4

Fig. 5.5 Swap mutation operation

5.4.6 Inversion

The main object of the inversion operator is to modify the relative positions of some the groups (i.e., the genes) in a given chromosome, so as to promote the succeeding crossover operations. In this application, the inversion operation is performed by overturning groups between two randomly selected inversion sites, however, subject to problem-specific hard constraints. Therefore, in the case that the resulting chromosome violates any specific constraints, a repair mechanism is employed.

Algorithm 4 The two-point inversion operator

Stage 1 With inversion probability p^i, select chromosome k for inversion operation

Stage 2 If k is selected, then randomly select two inversion sites i_1 and i_2

Stage 3 Overturn the groups between the sites i_1 and i_2

Stage 4 If any constraint is violated then repair the chromosome, otherwise continue

Stage 5 Repeat steps 1 through 4, until inversion is attempted on all the chromosomes in the population.

Figure 5.6 illustrates the two-point inversion procedure, based on an example of the chromosome $P_1 = \begin{bmatrix} 1 & 2 & 3 & 4 \end{bmatrix}$ with corresponding groups of items [{1, 3}, {4, 7}, {2, 6}, and {8, 5}].

The inversion mechanism overturns groups 2, 3, and 4 which lie between the two inversion sites, resulting in chromosome $\begin{bmatrix} 1 & 4 & 3 & 2 \end{bmatrix}$ with corresponding groups [{1, 3}, {8, 5}, {2, 6}, and {4,7}].

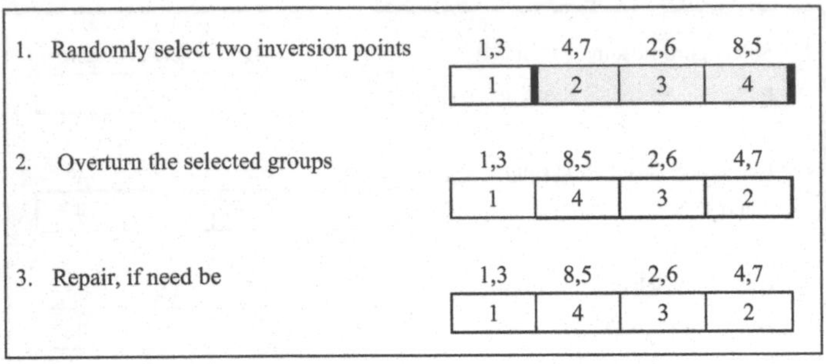

1. Randomly select two inversion points	1,3	4,7	2,6	8,5
	1	2	3	4

2. Overturn the selected groups	1,3	8,5	2,6	4,7
	1	4	3	2

3. Repair, if need be	1,3	8,5	2,6	4,7
	1	4	3	2

Fig. 5.6 Two-point inversion operator

5.4.7 Termination

The iterative algorithm searches the solution space continually until the termination condition is satisfied. In this case, the termination condition is defined by the user-defined maximum number of iterations (or generations), *maxgen*, which is selected according to expert opinion.

5.5   Experimental Tests and Results

In order to show the utility of the FGGA and its competitive advantages, three sets of experimental tests were considered. The first experiment involved a teaching group formation problem from a university. The second considered the comparative success rates of FGGA when tested 25 times on a known problem with a known solution. The third and final test consisted of a more extensive test with 20 generated problem sets. The algorithm was developed using Java 8, on a 1.70 GHz speed processor with 4 GB RAM. Following some preliminary experimentations, genetic parameters, including crossover, mutation, and inversion probabilities, were determined for the execution of the algorithm runs as listed in Table 5.1.

Table 5.1 Genetic parameters

Parameter	Value
Number of generations	500
Population size p	30
Initial crossover probability p_o^c	0.45
Initial mutation probability p_o^m	0.15
Initial inversion probability p_o^i	0.08

5.5.1 Experiment 1: Teaching Group Formation

First, the algorithm was tested on a problem concerned with formation of teaching groups in a university engineering department with 22 academic staff and 55 courses. Each of the staff has some level of knowledge about the available courses, and the knowledge ratings are estimated in the range [0, 5]. A rating of 5 would imply that the staff member has expert knowledge of the whole subject matter, while a rating of 0 would imply that the staff member has no specialized knowledge on that particular course. Based on the estimates of the ratings, a knowledge matrix was constructed. In retrospect, the objective is to arrange the staff into teaching groups, so that the overall knowledge of the courses in each group is as much satisfactory as possible.

For illustration purposes, the computational performance of the FGGA was compared with that of the basic GA which uses a conventional coding scheme where item numbers are represented as the positions in the chromosome and group numbers are coded as the genes of the chromosome, similar to the coding scheme used for group technology-based cellular manufacturing system design in Onwubolu and Mutingi (Onwubolu and Mutingi 2001). The basic genetic operators (crossover, mutation, and inversion) were used in the basic GA.

A transcription of the computational results is shown in Fig. 5.7. It is shown from the evolution search process depicted in the graph that the best solution for the FGGA is higher than that of GA in over most of the generations from 0 to 500. In addition, the final result for FGGA was 0.995, which is much higher than 0.935 obtained by the basic GA.

The next experiment compares FGGA with other competitive metaheuristic algorithms, that is, GA and particle swarm optimization (PSO).

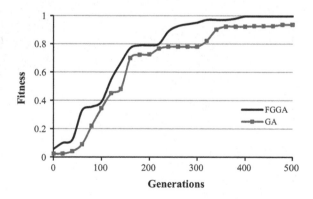

Fig. 5.7 Transaction of the comparative evolution process for FGGA and GA

5.5.2   Experiment 2: Comparative FGGA Success Rates

The second set of experiments compared the performance of FGGA with other competitive global optimization approaches; namely, the basic GA and PSO were used for the comparative analysis. The three algorithms were tested for search success rate based on a hypothetical problem consisting of 5 items and 15 resources with a known optimal solution. Each algorithm was run 25 times, while recording the search success rate and the average computation (CPU) times. Table 5.2 shows the results of the experiment. All the tested algorithms managed to obtain the desired optimal solution. However, in terms of search success rate, FGGA rated 100 %, followed by PSO rated 93.33 %, while GA rated 90 %. Moreover, FGGA rated the best in terms of the average computational times, with 12.8 s, compared to 16.32 and 23.16 for PSO and GA, respectively. Therefore, the FGGA is effective and efficient, when compared to other competitive algorithms. The algorithm has a high potential to perform excellently over large-scale problems.

5.5.3   Experiment 3: Further Extensive Computations

Tests for further experiments were generated at random by creating a knowledge incidence matrix K which is already block diagonalized and then adding. The

Table 5.2 Comparative analysis based on fitness and computational time

Problem set	Items	Resources	Group size	Best fitness	CPU time
1	20	30	3	1.000	12.9
			4	0.913	13.21
			5	0.781	14.27
			6	0.858	13.15
2	30	50	5	0.891	17.62
			6	0.692	26.79
			7	1.000	38.61
			8	0.925	34.13
3	40	80	7	0.802	38.76
			8	1.000	36.73
			9	0.894	47.34
			10	0.712	67.43
4	50	100	10	1.000	59.42
			12	0.732	62.15
			14	0.801	65.04
			15	0.767	81.18
			16	0.812	82.71

matrices were then perturbed to distort the diagonalized structure. This approach helped to ensure that the global optimum solutions were known a priori for testing the performance of the algorithm. Medium- to large-scale test problems were created and solved using FGGA. Of the entire set of problems, FGGA managed to accurately recover all the optimum solutions. Table 5.2 presents the solution fitness in terms of satisfaction level, μ, and the computational times in terms of CPU times.

The results show that the algorithm is highly capable of solving large-scale problems within a reasonable computation time, producing high-quality solutions. Therefore, the algorithm is effective and efficient for solving medium- to large-scale team formation problems. Furthermore, by using fuzzy set theoretic concepts, the algorithm can incorporate fuzzy management goals and aspirations, and the decision maker's choices and preferences, which all enhance decision making in a fuzzy environment. More realism is infused into the decision-making process through the use of weights in the presence of conflicting multiple criteria.

5.6 Summary

The performance of the entire organization is largely dependent on how well organized human resources are. Organizing human resources may entail formation of project teams, audit teams, specific task forces, multi-national work-force teams, research and development teams, and other forms of team-based work. The process of organizing human resources into teams is non-trivial. It has been realized that most decision makers in such areas are confronted with complexities related to the imprecise choices by human resources on their preferred team mates, skills ratings or qualities, management aspirations, and other multiple human resource-centered constraints and preferences. Due to the presence of fuzzy multiple optimization criteria, the problem tends to be highly complex. It might be desirable to maximize the total knowledge of individual staff in a team and to maximize the total knowledge of the team on an individual task. Consequently, the development of multi-criterion decision-making techniques is essential.

In this chapter, a multi-criterion fuzzy grouping genetic algorithm was proposed for the team formation problem. Imprecise preferences, decision maker's choices, management aspirations, and other preference constraints can be modeled conveniently using fuzzy set theoretic concepts built in fuzzy grouping genetic algorithm. Computational experiments and results were presented in this chapter, illustrating the usefulness of the proposed algorithm. The proposed algorithm is computationally efficient and effective in that it can produce competitive results in manageable computation times.

The problem presented in this chapter has inherent characteristics that are a lot similar to other real-world grouping problems such as reviewer grouping problem in the presence of multiple criteria and construction of audit teams. Therefore, the application of the proposed grouping algorithm can be extended to these and other related grouping problems.

References

Agustın-Blas LE, Salcedo-Sanz S, Ortiz-García EG, Portilla-Figueras A, Pérez-Bellido AM (2009) A hybrid grouping genetic algorithm for assigning students to preferred laboratory groups. Expert Syst Appl 36:7234–7241

Agustın-Blas LE, Salcedo-Sanz S, Ortiz-Garcıa EG, Portilla-Figueras A, Perez-Bellido AM, Jimenez-Fernandez S (2011) Team formation based on group technology: A hybrid grouping genetic algorithm approach. Comput Oper Res 38:484–495

Chen L (2001) Multi-objective design optimization based on satisfaction metrics. Engineering Optimization. 33:601–617

Delgado M, Herrera F, Verdegay JL, Vila MA (1993) Post optimality analysis on the membership functions of a fuzzy linear problem. Fuzzy Sets Syst 53:289–297

Dereli T, Baykasoglu A, Das GS (2007) Fuzzy quality-team formation for value added auditing: a case study. J Eng Tech Manage 24(4):366–394

Falkenauer E (1998) Genetic algorithms for grouping problems. Wiley, New York

Fitzpatrick EL, Askin RG (2005) Forming effective worker teams with multifunctional skill requirements. Comput Ind Eng 48:593–608

Goldberg DE (1989) Genetic Algorithm in Search, Optimization, and Machine Learning, Addison-Wesley, Reading, MA

Keeling KB, Brown EC, James TL (2007) Grouping efficiency measures and their impact on factory measures for the machine-part cell formation problem: a simulation study. Eng Appl Artif Intell 20:63–78

Kashan AH, Akbari AA, Ostadi B (2015) Grouping evolution strategies: An effective approach for grouping problems. Appl Math Model 39(9):2703–2720

Mutingi M, Mbohwa C (2016) Healthcare staff scheduling: emerging fuzzy optimization approaches. CRC Press, Taylor & Francis, New York

Onwubolu GC, Mutingi M (2001) A genetic algorithm approach to cellular manufacturing systems. Comput Ind Eng 39:125–144

Sakawa M (1993a) Fuzzy sets and interactive multi-objective optimization. Plenum Press, New York

Sakawa M (1993b) Fuzzy sets and interactive multi-objective optimization. Plenum Press, New York

Strnad D, Guid N (2010) A fuzzy-genetic decision support system for project team formation. Appl Soft Comput 10(4):1178–1187

Tavana M, Smither JW, Anderson RV (2007) D-side: a facility and workforce planning group multi-criteria decision support system for Johnson Space Center. Comput Oper Res 34(6):1646–1673

Tseng TL, Huang CC, Chu HW, Gung RR (2004) Novel approach to multi-functional project team formation. Int J Project Manage 22(2):147–159

Wi H, Mun J, Oh S, Jung M (2009a) Modeling and analysis of project team formation factors in a project-oriented virtual organization (ProVO). Expert Syst Appl 36(3):5775–5783

Wi H, Oh S, Mun J, Jung M (2009b) A team formation model based on knowledge and collaboration. Expert Syst Appl 36(5):9121–9134

Chapter 6
Grouping Learners for Cooperative Learning: Grouping Genetic Algorithm Approach

6.1 Introduction

Cooperative learning is a teaching technique in which small teams or groups of learners, with a different status or level of ability, use a variety of learning activities so as to enhance their understanding of a particular subject. This essentially involves organizing and grouping learners for improved academic and social learning experiences (Chen et al. 2012; Lynch 2010; Aldrich and Shimazoe 2010; Rubin et al 2011). Among other useful teaching techniques, cooperative learning is a very effective technique for effective learning in the classroom (Chen 2012; Chen et al. 2012; Tsay and Brady 2010; Brown and Ciuffetelli 2009; Scheurell 2010; Kose et al. 2010). Some of the expected outcomes of the technique are (i) academic gains such as improved learner's critical thinking and learner retention and (ii) improved self-esteem and student satisfaction, and positive social and race relations. In this view, it is important to implement cooperative learning strategies to create a fruitful learning environment that promotes the learning performance of students. However, in the real world, the implementation of the teaching strategy is non-trivial.

Organizing learners into effective cooperative groups is a complex grouping problem. This is due to complicating features and constraints. In the case of large classes, organizing learners into groups may be a challenge. Constraints relating to learner preferences, specialty, ability, gender issues, group size limit, social status, and other problem-specific restrictions. With varying group sizes and increasing class sizes, the grouping process becomes even more complex. The number of possible groupings increases exponentially. The optimal solution cannot be obtained in polynomial time. Conventional approaches, such as random grouping, selection by students themselves, selection by facilitator, and mathematical programming, may be time-consuming. Facilitators desire quick solution approaches that provide optimal or near-optimal solutions. In sum, the cooperative learners' grouping problem possesses the following complicating features:

© Springer International Publishing Switzerland 2017
M. Mutingi and C. Mbohwa, *Grouping Genetic Algorithms*,
Studies in Computational Intelligence 666,
DOI 10.1007/978-3-319-44394-2_6

1. The presence of a myriad of hard and soft constraints, which renders the problem highly restricted;
2. The complex grouping structure of the problem with academic and social interactions; and
3. The presence of numerous combinations that constitute the solution space of feasible groups of learners.

Therefore, developing effective and efficient approaches is essential. The purpose of this chapter is to present a global metaheuristic grouping algorithm for solving the grouping problem. By the end of the chapter, the reader should be able to do the following.

1. To define the cooperative learners' grouping problem, showing its complex grouping structure;
2. To analyze and model the grouping problem based on a metaheuristic grouping genetic approach; and
3. To perform computational experiments and evaluate the effectiveness and efficiency of the proposed grouping approach.

The rest of the chapter is organized as follows: The next section reviews related extant literature on the implementation approaches to cooperative learning. Section 6.3 explains the cooperative learners' grouping problem. Section 6.4 outlines the proposed GGA for the problem. Computational results and discussions are presented in Sect. 6.5. Conclusions and further research prospects are presented in Sect. 6.6.

6.2   Related Literature

Efficient grouping of learners into groups is the most challenging task. Conventional approaches exist in the literature. For instance, Johnson et al. (1994) outlined three approaches, namely random sampling, selected by teacher, and selected by students themselves. Chen (2012) and Chen et al. (2012) used a hybrid approach based on a combined sociometry and genetic algorithm to solve the complex cooperative learners' grouping problem. In trying to get the best solutions, one can use the relational network of learners (Chen 2012).

Moreno (1934) defined sociometry as the inquiry into the evolution and organization of groups and their relative position. It is useful for quantifying the social relationships in groups by exploring hidden structures that give the group its form, such as alliances, subgroups, hidden beliefs, and ideological agreements. In this vein, sociograms can be used to represent patterns of group interactions on the basis of various criteria, including social relations, channels of influence, and lines of communication (Moreno 1934; Jennings 1987; Cilessen 2011).

A number of indices have been defined to measure the social status of individuals (Chen 2012). In the development of sociometric methods over the years, the use of positive and negative nominations has been widely accepted (Cilessen 2011). A status is defined as a function of these two types of nominations, for instance, the difference between the two. One basic way to measure the social status of a learner is to use the relative or a normalized score known as *SSI* as follows:

$$\text{SSI} = \frac{N^{TC} - N^{TR}}{m - 1} \tag{6.1}$$

where N^{TC} is the total number of choices by others, N^{TR} denotes the total number of rejects by others, and m is the total number of learners. Since a learner cannot choose oneself, the expression $m - 1$ is used, instead of m.

The value SSI lies in the range $[-1,1]$, and the higher the value, the more the popularity of a learner within that particular group. Nevertheless, the index considers only one-way choice; that is, the mutual interaction between members from this index cannot be captured.

To consider the mutual choices between members, Ho (2002) defined the mutual choices between members, and ISSS is defined as follows:

$$\text{ISSS} = \frac{1}{2} \left(\frac{N^{TC} - N^{TR}}{m - 1} + \frac{N^{MC} - N^{MR}}{N^P} \right) \tag{6.2}$$

where N^{TC}, N^{TR}, N^{MC}, and N^{MR} denote the total choices by others, the total rejects by others, the mutual choices, and the mutual rejections, respectively; N^P denotes the maximum allowable number of choices by a learner.

The value ISSS is in the range $[-1,1]$, and the higher the value, the more the popularity of a learner within that particular group.

Although these approaches may be desirable to use and can provide solutions over small-scale problems, it is difficult to obtain optimal solutions, especially over large-scale problems with large classes.

6.3   Cooperative Learners' Grouping Problem

The cooperative learners' grouping problem is concerned with partitioning or clustering s students into g groups, where each group consists of n_g learners. Let $M = \{1, 2, \ldots, m\}$ be a set of learners, $G = \{1, 2, \ldots, g\}$ denote a set of groups of learners, $P = \{1, 2, \ldots, n_g - 1\}$ denote a set of learner's group mates, $C = \{1, 2, \ldots, c\}$ be a set of learners' choices, $Q = \{1, 2, \ldots, n_q\}$ be a set of grade levels, $R = \{1, 2, \ldots, n_r\}$ be a set of social status levels, and $T = \{1, 2, \ldots, n_t\}$ be a set of genders.

In this problem, learners are allowed to make their preferred group mates with whom they wish to cooperate. Learners' preferences are represented by indices 1, 2, $\ldots$, c, denoting a decreasing order of preferred choices. Then, let c_{ik} denote the preference of learner i on being grouped with partner k. That implies that if learner i did not select part k, then c_{ik} is assigned a large penalty number. The parameter c_{ki} is defined in a similar manner. Therefore, the problem can be formulated as follows:

$$\text{Min } Z = \sum_{i \in M} \sum_{j \in G} e_i x_{ij} \qquad (6.3)$$

$$\sum_{j \in G} x_{ij} = 1 \quad \forall \; i \in M \qquad (6.4)$$

$$\sum_{i \in M} x_{ij} = n \quad \forall \; j \in G \qquad (6.5)$$

$$\sum_{i \in M} x_{ij} z_{iq} = a_q \quad \forall \; j \in G, \; \forall \; q \in Q \qquad (6.6)$$

where

$$x_{ij} = \begin{cases} 1 & \text{If learner } i \text{ is assigned to group } j \\ 0 & \text{Otherwise} \end{cases} \qquad (6.7)$$

$$z_{iq} = \begin{cases} 1 & \text{If learner } i \text{ has an attribute } q \\ 0 & \text{Otherwise} \end{cases} \qquad (6.8)$$

$$e_i = \sum_{k \in G} \left(f(c_{ik}) + f(c_{ki}) \right) \quad \forall \; i \in M \qquad (6.9)$$

Expression (6.3) minimizes the total scoring value. Expression (6.4) ensures that each student is assigned to exactly one group, while (6.5) requires that each group g has n_g students. All learners' attributes are represented in (6.6), ensuring that exactly a_q learners with attribute q are represented in a group. Expression (6.9) represents a scoring function $f(\cdot)$ which is normally implemented as a square function.

The next section presents the proposed grouping genetic algorithm for the learners' grouping problem.

6.4 A Grouping Genetic Algorithm Approach

The proposed grouping genetic algorithm consists of a unique group coding scheme and grouping genetic operators, including group crossover, group mutation, and inversion. The application of the algorithm to the cooperative learners' grouping problem is presented in this section.

Learners: 1,3,5   4,7   2,6,8

Group: | 1 | 2 | 3 |

Fig. 6.1 The group encoding scheme

6.4.1 Group Encoding Scheme

For the cooperative learners' grouping problem, the items are the learners ($i = 1, 2, \ldots, m$), who are to be clustered into g groups ($j = 1, 2, \ldots, g$). Every learner i can be allocated to any group, subject to group size limits. This can be represented by a group encoding scheme. For illustration purposes, consider a class of 8 learners to be partitioned into g groups, where each group should have at least 2 members and at most 3. Figure 6.1 shows a chromosomal representation of a typical solution to this example, with groups $j = 1, 2$, and 3, corresponding to sets of learners {1,3,5}, {4,7}, and {2,6,8}, respectively.

It is important to note that group sizes may vary, but within a given range. In addition, each item (student) can only be found in only one group. This allows the genetic operators to work on groups, rather than items. In this respect, the GGA operators take advantage of the group structure of the problem, preserving the group spirit of the basic building blocks of the chromosome. One item taken singly has little or no positive influence on the global search process (Kashan et al. 2015; Falkenauer 1994, 1996, 1998).

6.4.2 Initialization

Selecting a good initialization procedure can significantly improve algorithm efficiency. For this reason, a constructive heuristic is used for generating the initial population. After accepting the input parameters, the algorithm randomly selects g out of m learners. Based on their mutual scores against the rest ($m - g$), each of the selected learners successively chooses a group member with the highest scores. The process is repeated until all the $m - g$ learners are included in groups. The initialization procedure is presented in Algorithm 1 below:

Algorithm 1 Initialization Algorithm

1. Input parameters: $maxgen, p^c, p^m, p^i, m, g$
2. Initialize, $x = 1$ //initialize population counter
3. Randomly select the first g learners
4. Let the g learners initiate g groups, ($j = 1, 2, \ldots, g$).
4. **While** ($x \leq m\text{-}g$) **do**
4.   **For** $i = 1$ to g
5.   Assign next available learner with highest mutual score to group G_i
6. $x = x + 1$ //increment population counter
7.   **End**
8. **End**

Following the initialization procedure, the population is evaluated and passed on to the selection and crossover operation. Evaluation is performed according to expression (6.3).

6.4.3   Selection and Crossover

Chromosomes are selected according to the rank-based wheel selection strategy, where candidates are sorted and ranked based on their relative quality (Gen et al. 2008; Holland 1975). For a population of size p, each candidate is assigned a rank R_k ($k = 1, ..., p$), where the best candidate is assigned the highest rank. Therefore, the fitness f_k associated with each candidate is given by:

$$f_k = 2R_k/(p(p+a))$$ (6.10)

The fitness values obtained are then associated with the intervals of the roulette wheel. This mechanism is used to select, with replacement, the parent chromosomes for crossover.

Following the selection procedure, a two-point crossover mechanism is applied to pairs of the selected chromosomes. The mechanism is explained in Algorithm 2 below;

Algorithm 2 Two-point crossover

1. Input: p^c, population P
2. **Repeat**
3.   Randomly select two parents, P_1 and P_2, $P_1 \neq P_2$;
4.   Select crossing sections for P_1 and P_2;
5.   Cross P_1 and P_2; interchange crossing sections;
6.   Obtain two offspring O_1 and O_2;
7.   Eliminate doubles, avoiding the crossed items;
8.   Insert misses; begin from where doubles were eliminated;
9.   Repair, if necessary;
10. **Until** (*poolsize* is reached)
11. **End**
12. **Return**

The crossover mechanism is repeated until the required population (*poolsize*) of new offspring is created.

Figure 6.2 illustrates the crossover operation. Assume that two parent chromosomes, $P_1 = [\{1,2\}\{4,9\}\{3,6,8\}\{5,7\}]$ and $P_2 = [\{8,2\}\{3,5\}\{6,9\}\{1,4,7\}]$, are randomly selected. A crossing section is selected for each parent, resulting in the crossing of the groups within the crossing sections. However, the offspring may

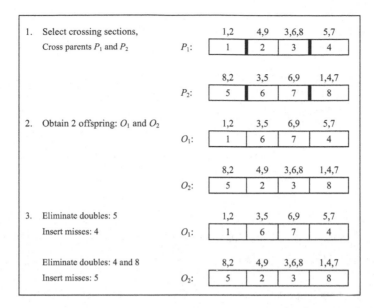

Fig. 6.2 The group crossover operation for cooperative learners' problem

have missing items (called *misses*) or repeated items (called *doubles*). The crossover process yields two offspring, O_1 and O_2, and both have misses and doubles which should be repaired. To remedy the offspring, doubles outside the crossing sections are knocked out, and misses are inserted beginning from outside the crossing sections. After the repair mechanism, the resulting offspring are $O_1 = [\{1,2\}\{3,5\}\{6,9\}\{5,7\}]$ and $O_2 = [\{8,2\}\{4,9\}\{3,6,8\}\{1,4,7\}]$.

After crossover, a new population (*newpop*) is created by combining the best performing offspring and the current population (*oldpop*).

6.4.4   Mutation

Mutation is primarily designed for exploitation through slight perturbations of the chromosomes at a low probability p^m, to improve current solutions (Gen and Cheng 2000;

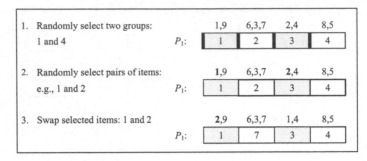

Fig. 6.3 The group mutation operation

Goldberg 1989). The operator works on groups rather than single items of the selected chromosomes. In this application, the basic swap mutation is employed.

Algorithm 3 Mutation operator

1. Input: p^m, population P
2. Select chromosome P_1 at a probability p^m;
3. Select from P_1, two different groups, g_1 and g_2;
4. **For** $i = 1$ to $size$ //$size$ = group size
5.   **If** $\{\text{prob}(p^m = \text{true})\}$**Then**
6.   Select item $i_1 = g_1[i]$ and $i_2 = g_2[i]$;
7.   Swap i_1 and i_2 selected items;
8.   **End**
9.   Repair chromosome P_1, if necessary.
10. **End**
11. **Return**

   In the case of any infeasible chromosomes due to violation of hard constraints, a repair mechanism is employed to restore the resulting chromosome. Figure 6.3 demonstrates the swap mutation mechanism using an example of a chromosome [1 2 3 4] with groups of items [{1,9}{6,3}{2,4,7}{8,5}].

   By selecting items from two selected groups (groups 1 and 3) and swapping them at a probability p^m, items 1 and 2 were successfully swapped. The process is repeated for the rest of the pairs of items, until all the pairs of items are tried for swapping.

6.4.5 Inversion

During the iterative search process, the population may prematurely converge toward a particular solution, which may limit the effectiveness of the crossover operator by which new regions of the solution space are explored. Diversity in the population is essential for exploration. However, after an appreciable number of iterations, the population should eventually converge to an optimal or a

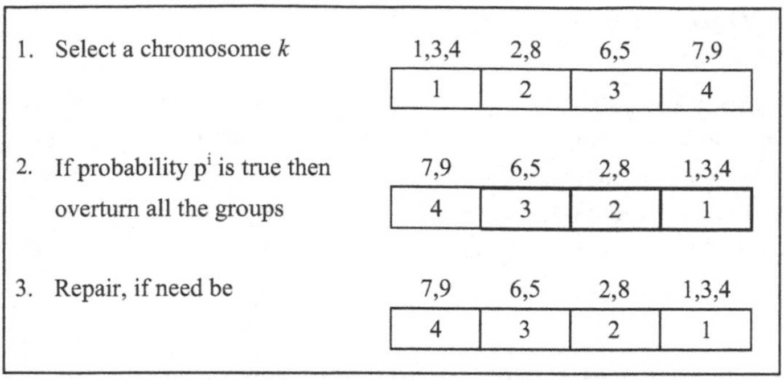

1. Select a chromosome k	1,3,4	2,8	6,5	7,9
	1	2	3	4

2. If probability p^i is true then overturn all the groups	7,9	6,5	2,8	1,3,4
	4	3	2	1

3. Repair, if need be	7,9	6,5	2,8	1,3,4
	4	3	2	1

Fig. 6.4 The full inversion operator for the learners' grouping problem

near-optimal solution. Therefore, a cautious trade-off between the need for exploration and the need for convergence is essential. A dynamic or adaptive control of diversity is a viable option. An adaptive inversion operator is employed in this GGA application. The algorithm for the inversion operator is as follows:

Algorithm 4 Full inversion operator

1. **Input**: Population P; inversion probability p^i;
2. **If** {prob p^i = true} **Then** perform inversion
3.   Overturn the entire chromosome
4.   Repair, if necessary
5. **End**
6. **Return**

For the purpose of illustration, an example of the inversion process is provided. Assume that a chromosome $k = [1\ 2\ 3\ 4]$ is selected for inversion, and the probability of inversion $prob\ (p^i)$ comes true. That implies that the whole chromosome k is overturned to $[4\ 3\ 2\ 1]$. The full inversion operation is illustrated in Fig. 6.4.

The resulting chromosome is tested for any violation of hard constraints, in which case the chromosome is repaired by a repair mechanism. However, we assume that no problem-specific constraints are violated in this problem.

For a dynamic inversion, the inversion operator is applied at a low probability, p, which decays over the iteration count;

$$p^i(t) = p_0^i e^{-\alpha(t/T)}$$

where α is an adjustment parameter in the range $[0,1]$; t is the iteration count; T is the maximum count; and p_0^i is the initial inversion probability. This allows the population to converge to a solution after the desired number of iteration counts is reached.

6.4.6 Termination

The algorithm is executed until a termination condition is reached. In this application, termination is controlled by the number of iterations (or generations). The maximum number of iterations, *maxgen*, is defined by the expert user. The next section presents computational experiments, comparative analysis results, and relevant discussions.

6.5 Computational Results and Discussions

The GGA was developed using Java 8, on a 1.70-GHz speed processor with 4-GB RAM. The algorithm was tested on a class of 80 students, varying the range of group sizes from 5 to 10. Each experiment was tested over 20 runs, and the best result among the trial runs was recorded as the solution. For the purpose of illustration, computational results and discussions are presented.

6.5.1 Preliminary Experiments

For preliminary experiments, the population size was varied from 10 to 40, while running the algorithm for 500 iterations and observing the best solution at each run. The variation of the best fitness value was recorded over 500 iterations. Moreover, preliminary settings for crossover, mutation, and inversion probabilities were 0.4, 0.1, and 0.07, respectively.

A transcription of the GGA search process is presented in Fig. 6.5. By varying the population size *p* from 10 to 40, it can be seen that the optimization search

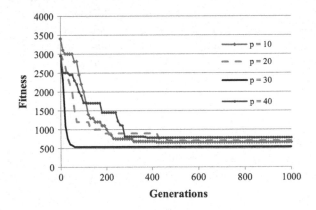

Fig. 6.5 A transcription of the GGA optimization search process

Table 6.1 Genetic parameters and their values

Parameter	Value
Population size	30
Crossover probability	0.45
Mutation probability	0.20
Inversion probability	0.10

process was best with $p = 30$. With population size set at $p = 30$, the fitness function settled to the near-optimal solution much faster than other settings.

In addition to the above experimental runs, crossover (p^c), mutation (p^m), and inversion rates (p^i) were varied and tested over 500 iterations, while observing the best fitness values. From these experiments, the selected genetic parameter values are determined in Table 6.1. The population size p and the maximum number of generations, *maxgen*, were set at 30 and 500, respectively. Furthermore, the crossover, mutation, and inversion probabilities were set at 0.45, 0.2, and 0.1, respectively.

After the preliminary experiments, the GGA was tested for its comparative efficiency and effectiveness.

6.6 Comparative Results: GGA and Other Approaches

The GGA performance was compared with the branch and bound approach (B&B) which is capable of obtaining optimal solutions. Table 6.2 summarizes the comparative results between GGA and branch and bound. The first two columns

Table 6.2 Comparative results between GGA and branch and bound method

No.	m	n	B&B		GGA	
			Best fitness	CPU time (s)	Best fitness	CPU time (s)
1.	6	2	38	0.00	38	2.07
2.	6	3	88	0.21	88	9.18
3.	8	2	68	12.60	68	31.73
4.	8	4	566	3.23	566	41.96
5.	12	2	N/A	980.22	178	46.37
6.	12	3	N/A	*	648	42.09
7.	15	3	N/A	*	988	66.78
8.	15	5	N/A	*	3020	71.35
9.	20	2	N/A	*	366	89.14
10.	20	4	N/A	*	460	121.88
11.	40	2	N/A	*	1320	189.34
12.	60	2	N/A	*	2636	266.82

*Computation time too long compared with GGA

Table 6.3 The influences of the group size on the GGA optimization process

Group size	Best solution	Average solution	CPU time (s)
2	783	890.78	54.12
3	1766	1860.10	68.57
4	2840	3132.73	99.36
5	4890	5189.11	121.3
6	7812	8142.31	138.9

describe the dimension of the problem in terms of the number of learners (m) and the size of a typical group (n). In the first 4 problems, the GGA procedure, like B&B, obtained optimal solutions within a few seconds. For the large-scale problems 5–12, B&B could not obtain solutions within the expected time.

The comparative results show that GGA is capable of getting optimal or near-optimal solutions with less computational times when compared to other methods.

6.6.1   Further Experiments

In this experiment, the number of items in each group was increased from 2 to 6, while keeping the number of choices constant. The aim is to demonstrate the effect of group size on the computational complexity, which directly influences computational time.

Table 6.3 highlights the variation of items in each group (n), the solution obtained, and the average computational or CPU time. It can be seen from this experiment that computational time increases with the increase in group sizes. This, in turn, implies that grouping complexity increases with the increase in group size.

6.7   Summary

The grouping genetic algorithm (GGA) is an effective metaheuristic algorithm that solves grouping problems by exploiting their group structures. Rather than working on individual genes (or items), the GGA operators, namely group crossover, group mutation, and group inversion, are designed to work on groups of genes, which gives the algorithm more effective and efficient search characteristics. By preserving the structure of the groups, which forms the basic building blocks of the algorithm, group similarity is maintained and improved with minimal disruptions, unlike when genetic operators work on single items.

Based on the group encoding scheme, GGA encodes the cooperative learners as items that are assigned to groups, according to mutual scores. The population of chromosomes is iteratively evolved over generations, exploring new regions of the solution space via group crossover operation and exploiting visited regions through group mutation. The versions of the crossover, mutation, and inversion operators used in this research are efficient and easy to apply. To enhance the search process, the inversion operator helps to dynamically maintain population diversity at acceptable levels, until final convergence as the iterations progress toward the termination condition.

Computational experiments show that the algorithm is efficient and effective, providing optimal or near-optimal solutions, even over large-scale problems. Further research application on closely related grouping problems, such as team formation and learners' grouping problem in a fuzzy environment, is highly recommended.

References

Aldrich H, Shimazoe J (2010) Group work can be gratifying: understanding and overcoming resistance to cooperative learning. College Teaching 58(2):52–57

Brown H, Ciuffetelli DC (eds) (2009) Foundational methods: understanding teaching and learning. Pearson Education, Toronto

Chen RC (2012) Grouping Optimization Based on Social Relationships. Math Prob Eng 1–19

Chen R-C, Chen S-Y, Fan J-Y, Chen Y-T (2012) Grouping partners for cooperative learning using genetic algorithm and social network analysis. In: The 2012 international workshop on information and electronics engineering (IWIEE). Procedia Eng 29:3888–3893

Cillessen AHN (2011) Sociometric methods. In: KH Rubin, WM Bukowski, B Laursen (eds) Handbook of peer interactions, relationships, and groups

Falkenauer E (1994) New representation and operators for GAs applied to grouping problems. Evol Comput 2(2):123–144

Falkenauer E (1996) A hybrid grouping genetic algorithm for bin packing. J Heuristics 2:5–30

Falkenauer E (1998) Genetic algorithms and grouping problems. Wiley, New York

Gen M, Cheng R (2000) Genetic algorithms and engineering optimization. Wiley, New York

Gen M, Cheng R, Lin L (2008) Network models and optimization: multi-objective genetic algorithm approach. Springer, Heidelberg

Goldberg DE (1989) Genetic algorithms: search, optimization and machine learning. Addison Wesley, Reading, MA

Ho CH (2002) A study of the effects on emotional experience, peer relationship and teacher-student relationship of the therapeutic approach to art activities in class in an elementary school setting. M.S. thesis, National Pingtung University of Education, Taiwan

Holland JH (1975) Adaptation in Natural and Artificial Systems. University of Michigan Press, Ann Arbor, MI

Jennings HH (1987) Sociometry in group relations, 2nd edn. Greenwood, Westport, Conn, USA

Johnson D, Johnson R, Holubec E (1994) Cooperative learning in the classroom. Association for Supervision and Curriculum Development, Alexandria, VA

Kashan AH, Akbari AA, Ostadi B (2015) Grouping evolution strategies: an effective approach for grouping problems. Appl Math Model 39(9):2703–2720

Kose S, Sahin A, Ergu A, Gezer K (2010) The effects of cooperative learning experience on eight grade students' achievement and attitude toward science. Education 131(1):169–180

Lynch D (2010) Application of online discussion and cooperative learning strategies to online and blended college courses. College Student J 44(3):777–784

Mitchell M (1996) An introduction to genetic algorithms. MIT Press, Cambridge

Moreno JL (1934) Who shall survive? A new approach to problem of human interrelations. Nervous and Mental Disease Publishing, Washington, DC

Rubin KH, Bukowski WM, Laursen B (eds) (2011) Handbook of peer interactions, relationships, and groups

Scheurell S (2010) Virtual warrenshburg: using cooperative learning and the internet in the social studies classroom. Soc Stud 101(5):194–199

Tsay M, Brady M (2010) A case study of cooperative learning and communication pedagogy: does working in teams make a difference? J Scholarsh Teach Learn 10(2):78–89

Chapter 7
Optimizing Order Batching in Order Picking Systems: Hybrid Grouping Genetic Algorithm

7.1 Introduction

Order picking is a crucial operational activity that is common in distribution warehouse systems. The activity is concerned with the retrieval of items from their respective storage locations in a warehouse so as to satisfy customer demands and expectations. Oftentimes, customer orders come in small volumes of various item types (or articles); this obviously complicates the planning and scheduling of the item retrieval operations. In such industrial situations, providing high-quality services to customers is a challenge to distribution warehouse systems. Decision makers have to develop decision support systems to facilitate their order picking decision-making process in the shortest possible times, so as to ensure satisfactory customer service in terms of customer waiting times, processing times, delivery times, and other pertinent performance criteria.

From a more general perspective, order picking systems can be divided into two broad categories (Wascher 2004; Henn and Wascher 2012). These are described as follows:

1. Pickers-to-parts systems: Order pickers search through the warehouse to pick items according to customer orders; and
2. Parts-to-picker systems: Automated storage retrieval vehicles deliver the requested items to order pickers (usually stationary order pickers).

Further to the above perspectives, three operational planning problems can be identified from the first category, as follows (Caron et al. 1998):

1. Article location problem: This problem is concerned with the assignment of articles or items to specific storage locations;
2. Order batching problem: This pertains to the problem of grouping customer orders into cost-effective batches that meet customer demands; and
3. Picker routing problem: The problem involves the scheduling and routing of order pickers through the warehouse

© Springer International Publishing Switzerland 2017
M. Mutingi and C. Mbohwa, *Grouping Genetic Algorithms*,
Studies in Computational Intelligence 666,
DOI 10.1007/978-3-319-44394-2_7

Of the three categories highlighted above, this chapter is concerned with the order batching problem, which is very crucial for efficient operations in manual order picking systems in distribution warehouses. Thus, order batching is the method of grouping a set of orders into a number of subsets; each of which can be retrieved by a single picking tour. Quite pivotal in enhancing the efficiency of warehouse operations, order batching is a highly repetitive material-handling process that often demands large-scale manual labor input, in the form of human operators. In addition, the order batching problem has a substantial influence on other related planning activities such as warehouse design, article location, and picker routing. As such, it is essential to optimize order batching processes.

In most warehouse systems, large volumes of items are stored according to their various types and forms. Oftentimes, a number of customers tend to demand small volumes of various item types, which ultimately leads to numerous order picking movements in the warehouse. As a result, the operational costs may be too high due to long cumulative distance travelled during order picking process, labor costs, and other additional operational costs (Ho and Tseng 2006; Jarvis and McDowell 1991; Petersen 1997; Tompkins et al. 2003; Yu and de Koster 2009). Therefore, the order picking process should be as efficient and effective as possible, if the warehouse system is to remain competitive. In a competitive business world, it is essential to ensure high quality of service, so that customers are always satisfied. The challenge is to set up an efficient and effective order picking system that can provide an acceptable level of quality of service to customers.

Order batching is a complex combinatorial problem that is known to demand considerable computational resources (Henn et al. 2010; Henn 2012; Chen et al. 2005; Elsayed 1981; Elsayed and Unal 1989). Though iterative heuristic search algorithms have been proposed for the order batching problem, the solution process is often too time-consuming for most real-world order picking problem settings. In practice, solutions have to be generated on real time or at least within as short a period as possible. Interestingly, the concept of batching, or grouping customer orders, opens up the idea of grouping items according to desired criteria opens. This suggests that it is possible to model the problem in a more computationally effective and efficient way, by exploiting the problem's group structure. Since the problem is highly complex and combinatorial in nature, hybrid solution approaches that incorporate unique grouping genetic operators, constructive heuristics, and other heuristic algorithms are quite promising.

This chapter focuses on the development of a hybrid grouping genetic algorithm (HGGA) that makes use of unique grouping operators, constructive heuristics, and other heuristics. The envisaged learning outcomes for this chapter are as follows:

1. To understand the order batching problem and its associated computational challenges;
2. To understand how to computationally exploit the grouping structure of the problem; and
3. To develop a hybrid grouping genetic algorithm and to evaluate the computational performance of the algorithm.

Extensive numerical experiments are conducted to test the utility of the suggested computational procedure. The performance of the algorithm is compared against other benchmark heuristics.

The remainder of the chapter is as follow: Sect. 7.2 describes the order batching problem. Section 7.3 provides a review of the related literature. Section 7.4 proposes a hybrid grouping genetic algorithm for the order batching problem. Computational experiments are described in Sect. 7.5. Computational results and discussions are presented in Sect. 7.6. Finally, concluding remarks and further research are outlined in Sect. 7.7.

7.2   Order Batching Problem

To enhance understanding of the discussions that follow, a more comprehensive description of the order batching problem and its mathematical formulation is presented in this section.

7.2.1   Description of the Problem

In order picking systems, a customer order is defined as a set of non-empty order lines. Each order line is made up of a particular article and a specification of the required number of items. On the other hand, a pick list is a set of order lines that the order picker follows when retrieving items from the warehouse. It follows that the pick list consists of a group of order lines that must be retrieved together at a particular point in time. Therefore, the order lines in a picking list constitute a picking order which comes in two forms:

1. Pick-by-order: where the order lines in the list are meant to meet the demand for a single customer; and
2. Pick-by-batch: where the order lines in the list are meant to satisfy a group of customer orders.

In the real world, a picking order is treated as a batch, consisting of items that must be picked in a particular sequence as determined by a specific touring or routing strategy that the order picker should follow. Usually, the tour begins from a depot, proceeds to a series of respective storage locations, and returns to the depot.

Figure 7.1 shows a traversal of a typical warehouse based on a conventional heuristic routing strategy, represented by dotted lines. The shaded rectangle and squares represent the depot items to be picked, respectively. Every aisle containing a requested item is traversed entirely, subject to the capacity of the order picker. The aim is to minimize the total picking time, consisting of setup times for the picker tours, the travel time from the depot to item locations and back to the depot, the

Fig. 7.1 A traversal of a
warehouse based on a
heuristic

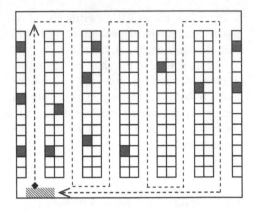

item search times, and the time to pick items, of which the travel time is the major concern. This translates to minimizing the total length of picker tours.

In summary, the order batching problem (OBP) can be described as follows (Wascher 2004): Given the article locations in a warehouse distribution system, the capacity of the picking device, and the customer orders to be picked, how best can the customer orders be grouped into batches so that the total tour length required to pick the requested items is minimized.

7.2.2 Problem Formulation

The OBP can be formulated as an integer programming model (Henn and Wascher 2012; Gademann and van de Velde 2005). It is assumed that all the feasible (not exceeding the capacity of the picking device) are identifiable. For ease of deliberation, the following notation is defined for the problem.

Notation:

n the number of customer orders

J set for customer orders, $J = \{1, \ldots, n\}$

C capacity of the picking device

c_j capacity required for order j ($j \in J$)

a_j ($a_{i1}, \ldots, a_{in}$), a vector indicating whether order j is in batch i ($a_{ij} = 1$) or not ($a_{ij} = 0$)

I set of all feasible batches

l length of each pricking tour

x_i 1 if batch i is chosen; 0, otherwise

Assume that $c_j \leq C$ ($j \in J$), and $\sum_{j=1}^{n} c_j \geq C$; otherwise, the problem will be trivial. The definition of the set I signifies that the picking capacity should never be violated, that is,

$$\sum_{j \in J} c_j a_{ij} \le C \tag{7.1}$$

Therefore, the set I can be generated from C and c_j $(j \in J)$. The length l_i of each pricking tour i $(i \in I)$ is determined from the set of all the orders included in the tour. If x_i is a variable indicating whether a batch is chosen or not, then the following model holds for the OBP;

$$\text{Min } z = \sum_{i \in I} l_i x_i \tag{7.2}$$

$$\text{sub to } \sum_{i \in I} a_{ij} x_i = 1 \quad \forall \; j \in J \tag{7.3}$$

$$x_i \in \{0, 1\} \quad \forall \; i \in I \tag{7.4}$$

Constraints (7.3) and (7.4) ensure that each customer order is found in exactly one of the selected batches. It can be seen that the OBP is an NP-hard in its strongest sense, if the number of customer orders in each batch is greater than two, which potentially leads to computational problems. As the number of customer orders increases, the number of binary variables increases exponentially (Henn and Wascher 2012; Gademann and van de Velde 2005).

7.3   Related Solution Approaches

A number of solution approaches exist in the literature (Hall 1993; De Koster et al. 2007; Henn et al. 2012; Henn and Wascher 2012). The approaches can be classified into four categories: (i) routing heuristics, (ii) mathematical programming techniques, (iii) constructive heuristics, and (iv) metaheuristics. A hybrid of these approaches may be favorable.

7.3.1   Routing Heuristics

A number of heuristic routing strategies have been proposed for the routing order pickers in a warehouse (Koster et al. 2007; Henn and Wascher 2012). One attractive advantage of these strategies is that they are computationally inexpensive.

Figure 7.2 shows four examples of conventional heuristic routing strategies, namely (a) the s-shape or traversal heuristic, (b) the midpoint heuristic, (c) the return heuristic, and (d) the largest gap heuristic routing strategies. The shaded squares represent the locations of the articles to be picked, while the dotted lines

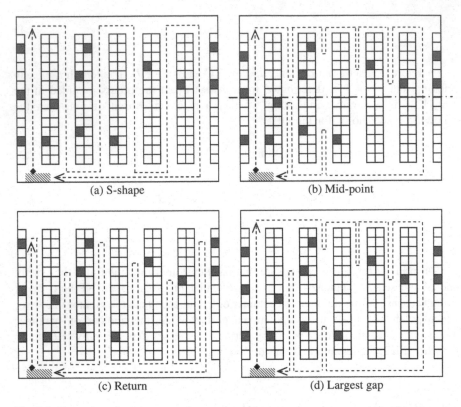

Fig. 7.2 Examples of routing strategies

represent the tours that the order picker has to follow when traversing through the warehouse.

According to the s-shape heuristic, the order has to enter and completely traverse any aisle that contains a requested item, and then proceeds to the next aisle containing a requested item. This continues until the front cross-aisle where the order picker has to pick the requested item(s) and return to the depot. The midpoint heuristic divides the warehouse into two areas, where picks in the front half are accessed from the front cross-aisle and those in the back half are accessed from the back cross-aisle. The largest gap heuristic emulates the midpoint strategy except that an order picker enters an aisle as far as the largest gap within an aisle, instead of the midpoint. Thus, the gap refers to the separation between any two adjacent picks, between the first pick and the front aisle, or between the last pick and the back aisle. In this respect, the largest gap is therefore the portion of the aisle that is not traversed. The heuristic strategies demonstrate the advantages of combining customer orders in one tour, as opposed to picking by single customer orders. Further, a composite heuristic requires that aisles with picks be traversed entirely or entered and left at the same end. This may be accomplished using dynamic programming (Roodbergen and De Koster 2001a, b).

7.3.2 Mathematical Programming Techniques

A few extant approaches have been suggested for providing optimal or near-optimal solutions to the OBP problems. Mathematical programming approaches include linear programming, integer programming, mixed integer programming, and dynamic programming. Gademann and Velde (2005) formulated the problem as a integer program and applied a branch-and-price algorithm with column generation. The approach provided optimal solutions for small-scale problems of up to 32 customer orders. Similarly, Bozer and Kile (2008) proposed a mixed integer program, providing near-optimal solutions for small problems of up to 25 customer orders. However, in the case of large problems, which are commonplace in industry, the application of constructive heuristics is a viable option in specific problem situations.

7.3.3 Constructive Heuristics

Constructive heuristics include priority rule-based algorithms, seed algorithms, savings algorithms, and data mining approaches (Wascher 2004). For instance, priority rule-based heuristics assign priorities to customer orders and then allocate to specific batches accordingly (Gibson and Sharp 1992). In seed algorithms, one customer order is chosen as a seed, and additional orders are generated and added according to given rules (Ho et al. 2008). In the savings algorithm, the potential savings obtainable from each possible combination of customer orders is evaluated each time the customer orders are allocated to a batch (Hwang and Kim 2005). In data mining and integer programming techniques, similarities of orders are determined based on predetermined association rules, so that the overall correlation between the customer orders is maximized (Chen and Wu 2005). Customer orders are then assigned to batches heuristically.

7.3.4 Metaheuristics

Several researchers proposed various metaheuristic approaches for the OBP problem situation (Hsu et al. 2005; Gademann and Velde 2005; Koster et al. 2007; Tsai et al. 2008; Albareda-Sambola et al. 2009; Henn et al. 2010; Henn and Wascher 2012). For instance, Gademann and Velde (2005) suggested an iterated improvement algorithm which begins with a scene generated by a first-come, first-served rule. Solution improvement is achieved through swap moves. In the same vein, Hsu et al. (2005) proposed a genetic algorithm approach for the OBP. Similarly, Tsai et al. (2008) modeled the batching and routing problems based on genetic algorithm. Henn and Wascher (2012) presented two approaches, based on the concepts of the tabu search metaheuristic. In general, metaheuristics are capable of providing near-optimal solutions with reasonable computation times. Other related approaches exist in the literature (Koster et al. 2007; Henn et al. 2012).

In addition to the solution approaches highlighted above, hybrid approaches have a great potential to perform better than known approaches. The next section presents a hybrid grouping genetic algorithm for the OBP.

7.4 Hybrid Grouping Genetic Algorithm for Order Batching

Unlike the conventional item-based genetic algorithm, the proposed HGGA is designed to build upon constructive heuristics and a grouping algorithm whose coding scheme and genetic operators are sensitive to the group features of the problem. HGGA utilizes a group coding scheme and constructive heuristics to generate and iteratively improve a population of solutions by looping through grouping genetic operators, that is, crossover, mutation, and inversion.

7.4.1 Group Encoding Scheme

HGGA codes batches instead of individual customer orders. As such, the length of the chromosomes may vary. Customer orders, called items ($i = 1, 2, \ldots, m$), are clustered into g groups ($j = 1, 2, \ldots, g$), where each group represents a batch. Every order (or item) i can be allocated to any group, subject to constraints. Figure 7.3 shows an example of the group coding scheme for a typical solution, with 11 customer orders partitioned into 5 groups. The chromosome encoding represents groups $j = 1, 2, \ldots, 5$, corresponding to sets of items {3,8}, {2,5,11}, {4,9}, {6,10}, and {7,1}, respectively. Thus, orders 3 and 8 are in group 1, orders 2, 5, and 11 are in group 2, and so on. Each item can only appear in one group.

The HGGA operators work on groups, rather than on single items of the chromosomes. In this respect, the algorithm tends to preserve important information coded in the building blocks of the group scheme of the chromosome (Kashan et al. 2015).

7.4.2 Initialization

The initialization procedure is responsible for accepting input parameters and generating the initial population of chromosomes. To improve the efficiency of the

Fig. 7.3 Group encoding scheme for order batching

Exams:	3,8	2,5,11	4,9	6,10	7,1
Group:	1	2	3	4	5

subsequent HGGA iterative procedure, the initial population of chromosomes is created using a constructive heuristic which ensures that only feasible solutions are created.

The algorithm calculates the savings list L by calculating the saving $s_{i,j}$ for each pair (i,j) based on the formula $s_{i,j} = s_{i,j} + s_{0,j} - s_{i,j}$. The list L is then sorted in descending order of the savings. Subsequently, the pairs of orders are sequentially considered and merged into groups, provided grouping and other capacity constraints are not violated. The process is repeated until the required number of groups is generated.

The pseudo-code for the initialization procedure, enhanced with the savings algorithm, is as shown in the algorithm:

Algorithm 1 Savings Constructive Initialization Algorithm

1. **Input:** parameters: *maxgen*, p^c, p^m, p^i, m, g, p
2. **Input:** data: set of customer orders, cost matrix
3. Initialize groups as pairs   // calculate savings list
4. **For** {every pair $\{i,j\}$ customer orders, $i \neq j$} **do**
5.   Calculate savings $s_{i,j} = s_{i,j} + s_{0,j} - s_{i,j}$
6.   Store $s_{i,j}$ in savings list L
7. **End**
8. Sort L in descending order
9. **While** $\{n < p\}$ **Do**   // Generate population of chromosomes
10. // check every entry in savings list L and merge if constraints are satisfied
11.   **For** {every entry $\{i,j\}$ in L} **Do**
12.   **IF** {group with order i and group with order j can be merge} **Then**
13.   merge the groups
14.   **End**
15.   **End**
16.   Count the current number n of groups
17. **End**

After the initialization process, the population is passed on to the iterative loop comprising selection, crossover, mutation, and inversion operations.

7.4.3   Selection and Crossover

HGGA uses a rank-based roulette wheel procedure to select two candidate parents for crossover operation. Each chromosome is assigned a portion of an imaginary roulette wheel, according to its rank-based fitness value. The fitness of each chromosome is evaluated in terms of total tour length, and the fitter ones have higher rank-based fitness values. It follows that fitter chromosomes receive relatively higher proportion of the roulette wheel.

Crossover is the most important of the group operators of the HGGA. By combining the genetic information in the candidate parents, new regions of the search space are explored, yielding new and better solutions. The group crossover procedure avoids disruption of the group structure of the chromosomes, which serves to preserve essential information encoded in the chromosomes.

Figure 7.4 presents an illustrative example of the crossover mechanism using two randomly selected chromosomes, $P1 = [1\ 2\ 3]$ with groups of items $\{1,3\}$, $\{4,7\}$, and $\{2,5,6\}$ and $P2 = [4\ 5\ 6]$ with groups $\{1,7,2\}$, $\{3,5\}$, and $\{6,4\}$. The crossover operation yields chromosomes $O1 = [1\ 6\ 3]$ containing groups of items $\{1,3\}$, $\{6,4\}$, and $\{2,5,7\}$, and $O2 = [4\ 5\ 2]$ containing groups of items $\{1,6,2\}$, $\{3,5\}$, and $\{4,7\}$.

The process is repeated until a population of new offspring (called *spool*) of size $= p^c \times p$ is created. The pseudo-code for the crossover operator, deriving from the description above, is presented in Algorithm 2.

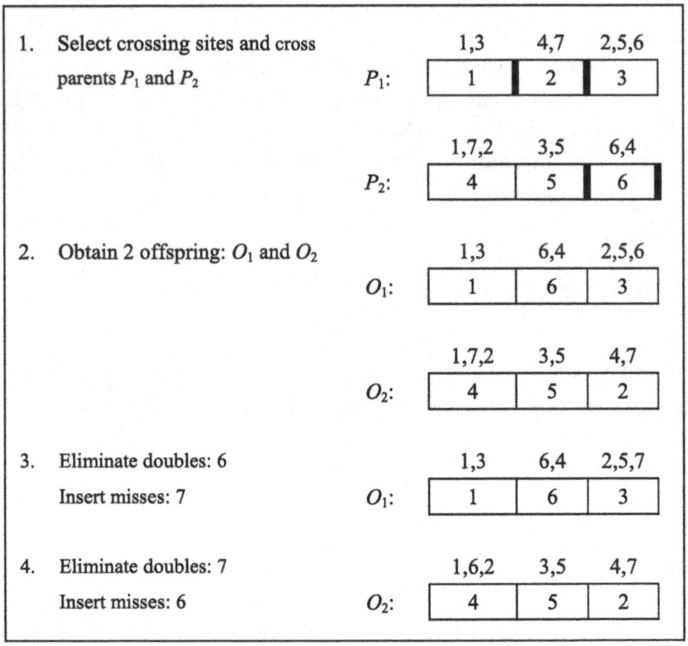

			1,3	4,7	2,5,6
1.	Select crossing sites and cross parents P_1 and P_2	P_1:	1	2	3
			1,7,2	3,5	6,4
		P_2:	4	5	6
			1,3	6,4	2,5,6
2.	Obtain 2 offspring: O_1 and O_2	O_1:	1	6	3
			1,7,2	3,5	4,7
		O_2:	4	5	2
			1,3	6,4	2,5,7
3.	Eliminate doubles: 6 Insert misses: 7	O_1:	1	6	3
			1,6,2	3,5	4,7
4.	Eliminate doubles: 7 Insert misses: 6	O_2:	4	5	2

Fig. 7.4 Group crossover operation for order batching

Algorithm 2 Group Selection and Crossover

Input: population p; probability p^c
1. Initialize count $size = 0$
2. **While** $\{size \leq$ desired $s\}$ **do**
3.   select parents P_1 and P_2
4.   select crossover points
5.   **IF** $\{prob(p^c = $ true$)$ **Then**
6.   cross groups defined by crossover points; obtain offspring O_1 and O_2
7.   **IF** $\{$any *doubles* or *misses*$\}$ **Then**
8.   eliminate doubles;
9.   insert missing items;
10.   **End**
11. $size = size + 2$;   // increment selectin pool counter
12.   **End**
13. **End**
14. **Return**

After crossover operation, a new population (called *newpop*) is created by combining best performing chromosomes from the selection pool *spool* and the old population *oldpop*. The new population is then passed on to the mutation operation.

7.4.4 Mutation with Constructive Insertion

Group mutation with constructive insertion exploits the neighborhood of candidate solutions in the current population. Small perturbations are performed on randomly selected chromosomes, at a low probability p^m. To improve the mutation procedure, a constructive insertion procedure is proposed. An example of the insertion procedure is given in Fig. 7.5:

Stage 1. Randomly choose two groups, for instance, first group 2 and second group 4, and delete the first group.
Stage 2. Create a new group in place of the deleted group, for instance, 2', and place it between the second group and its adjacent group, for example, groups 3 and 4.
Stage 3. Insert deleted items into other groups, according to similarity rules, while considering grouping and capacity constraints. Items with high similarity are grouped together according to similarity measures known a priori. Assume that there is a high similarity between 1, 3, and 5, and that all constraints are satisfied.

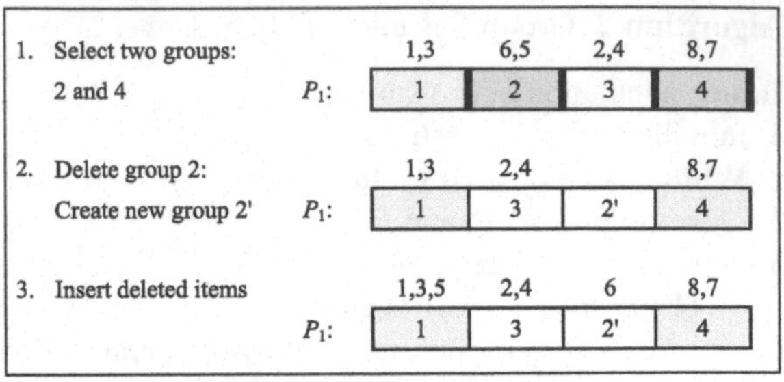

Fig. 7.5 Procedure for group mutation with constructive insertion

			1,3	6,5	2,4	8,7
1.	Select two groups: 2 and 4	P_1:	1	2	3	4
			1,3	2,4		8,7
2.	Delete group 2: Create new group 2'	P_1:	1	3	2'	4
			1,3,5	2,4	6	8,7
3.	Insert deleted items	P_1:	1	3	2'	4

With a probability of p^m, the mutation operation is performed over the entire population. A pseudo-code for the group mutation mechanism is presented in Algorithm 3

Algorithm 3 Mutation with Constructive Insertion

1. **Input:** population; population size p; mutation probability p^m
2. **For** {*count* = 1 to p} **Do**
3.   **If** {prob (p^m = true)} **Then**   // mutate at a low probability
4.   Randomly choose groups A and B; delete first group A
6.   Create new group A'; place it between B and its adjacent group B'.
7.   **For** {each chromosome} **Do**
8.   **If** {constraints satisfied} **Then**
9.   re-insert group A items by similarity measure
10.   **End**
11.   *count* = *count* + 1
12.   **End**
13.   **End If**
14. **End**

After mutation to an acceptable level of diversity, the population passes to the inversion operator.

7.4.5 Inversion

It is possible that as HGGA iterations progress, the population may converge prematurely, which potentially limits the effectiveness of the crossover operator. The inversion operator curbs this effect by overturning groups between two randomly selected inversion points. This is also essential for maintaining population diversity to an acceptable level. Figure 7.6 provides an example of inversion procedure; groups 2, 3, and 4 between the two inversion sites are overturned. As a result, chromosome [1 2 3 4] is transformed to [1 4 3 2].

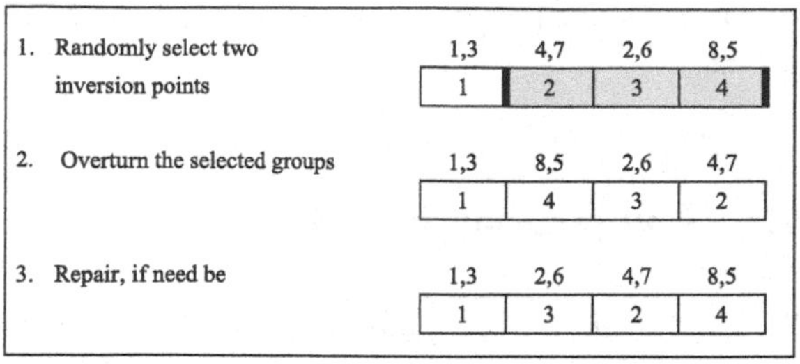

Fig. 7.6 Group inversion operation for the order batching problem

1.	Randomly select two inversion points	1,3	4,7	2,6	8,5
		1	2	3	4
2.	Overturn the selected groups	1,3	8,5	2,6	4,7
		1	4	3	2
3.	Repair, if need be	1,3	2,6	4,7	8,5
		1	3	2	4

The inversion operator is applied at a low probability, $p^i(t)$, which decays as the number of iterations increases,

$$p^i(t) = p_0^i \cdot e^{-\alpha(t/T)} \qquad (7.5)$$

where t is the iteration count; α is a constant in the range [0,1]; T is the maximum number of iterations; and p_0^i is the initial user-defined inversion probability. The resulting inversion mechanism is outlined in Algorithm 4.

Algorithm 4 Two-Point inversion

1. Input: population P; population size p, inversion probability p^i
2. **For** $i = 1$ to p **Do**
3.   **If** {prob $(p^i(t))$ = true} **Then**
4.   Randomly select two inversion points;
5.   Overturn the selected groups;
6.   Repair, if need be;
7.   **End**
8. **End**
9. **Return**

Computational experiments are presented in the next section, together with comparative results and discussions.

7.4.6 Termination

Termination condition is crucial for the algorithm to obtain the best possible population of alternative solutions. The proposed HGGA procedure is terminated when either or both of the following criteria are satisfied:

1. When the user-defined maximum number of generations or iterations, T, is reached, or
2. When a certain number of iterations are reached without any significant change in the current best solution.

Computational experiments, results, and discussions are presented in the next section.

7.5   Computation Experiments

To test the utility of the proposed HGGA solution approach, assume a single-block warehouse with two cross-aisles, a single depot at its corner, similar to the structure in Fig. 7.7. The structure is commonly used in order picking experiments (Henn et al. 2010; Petersen and Aase 2004; Petersen and Schmenner 1999; Henn and Wascher 2012). Further, the warehouse has 10 vertically oriented picking aisles, each with a capacity of 90 locations, which sums up to 900 storage locations in all. Each storage location is one unit length (UL). A center-to-center distance between two aisles amounts to 5 unit lengths 9 UL. The depot is 1.5 UL from the first storage locations on the extreme left aisle of the warehouse.

Each article in the warehouse occupies one storage location. The articles are distributed following a class-based storage assignment policy: (a) Class A consists of articles with high demand frequency; (b) class B consists of articles with average demand frequency; and (c) class C consists of articles with the lowest demand frequency. In this respect, A articles are located at the leftmost aisles, B articles are at the center aisles, and C articles are at the extreme right aisles. Moreover, assume that the demand frequencies of classes A, B, and C are 52, 36, and 12 %, respectively. This problem setups are common in real-world warehouses (Henn et al. 2010; de Koster et al. 1999a, b).

Customer orders are assumed to vary from 20 to 60 in steps of magnitude of 10 orders. The number of articles in each order is assumed to be uniformly distributed

Fig. 7.7 An exemplary
single-block warehouse

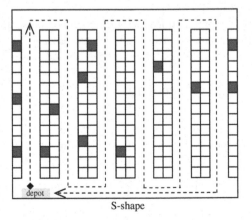

S-shape

Table 7.1 A summary of the problem characteristics

No.	Problem characteristic	Specific information
1.	Number of customer orders	20, 30, 40, 50, 60
2.	Demand frequency classes	A (52 %), B (36 %), C (12 %)
3.	Order picking device capacity	30, 45, 60, 75
4.	Articles in each order	Uniform [5,25]
5.	Routing strategy	S-shape

in [5, 6, ..., 25]. The sizes of the picking devices are 30, 45, 60, and 75. The S-shape strategy is assumed in this analysis. A summary of the problem characteristics is presented in Table 7.1.

From the above information, it can be seen that 20 different problem classes can be created for experimental analysis.

7.6 Computational Results and Discussions

Computational results reveal the efficiency and effectiveness of the proposed HGGA approach, together with how its performance compares with related algorithms. In this case, HGGA performance is compared with GA with item-centered encoding scheme.

7.6.1 Preliminary Experiments

Preliminary runs were done to determine the values of the genetic operators, that is, crossover probability (p^c), mutation probability (p^m), and inversion probability (p^i). The values of these parameters were varied in their respective experimental ranges in [0,1], while observing the behavior of the fitness function. In addition, the value of the population size p was varied from 10 to 50, in steps of magnitude of 10. Table 7.2 shows the final parameter values that were selected for both the HGGA and the basic GA.

Table 7.2 Selected genetic parameters and their values

Parameter	Experimental range	Selected value
Population size	10–50	30
Crossover probability	0.3–0.8	0.45
Mutation probability	0.05–0.3	0.20
Inversion probability	0.02–0.1	0.08

The termination criterion depended on the maximum number of generations which was set at 200. Further experiments were performed based on these parameter settings.

7.6.2 Further Experiments

To demonstrate the efficiency and effectiveness of the optimization search process of the HGGA, the performance of the algorithm was tested and compared to that of the conventional (basic) GA with item-centered encoding. Computational results are presented.

7.6.2.1 Solution Quality Over Iterations

Transcription of the evolutionary search process of the algorithm is given in Fig. 7.8. The evolutionary computation process illustrates the variation of the quality of the solution in terms of deviation from the best value over 200 generations (or iterations), that is, at the end of the evolutionary process. Four problem instances are considered, that is, (a) $n = 20$, $c = 30$, (b) $n = 30$, $c = 30$, (c) $n = 50$,

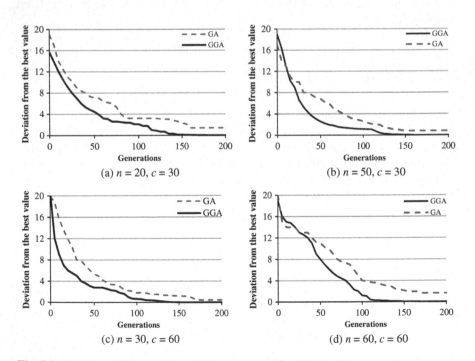

Fig. 7.8 Solution quality over the evolutionary process for 200 iterations

$c = 30$, and (d) $n = 60$, $c = 60$. The vertical axis denotes the deviation from the best value, while the horizontal axis shows the generations (iterations).

Key observations are realized from this analysis. GGA requires fewer generations to obtain near-optimal solutions when compared to GA. Typically, as can be seen from (a) to (c), good solutions are obtained within 140 generations. Furthermore, the quality of solutions obtained by GGA is much higher than the ones from the basic GA. Therefore, the GGA approach can provide near-optimal solutions with less computational resources. This can be attributed to the incorporation of constructive heuristics and the group-based coding structure, together with the unique grouping genetic operators (crossover, mutation, and inversion).

7.6.2.2   Comparative Computational Efficiency

Further to the above analysis, the efficiency of the basic GA and HGGA was compared in terms of computing times. It can be realized that the computational complexity of the problem is largely influenced by the number of customer orders, n, and the capacity of the order picker c. This implies that computing times are directly proportional to n and c. Table 7.3 presents the results of a comparative analysis between the performance of GA and HGGA in terms of computational (or CPU) times.

Table 7.3 Comparative computational times between basic GA and HGGA

n	c	GA (sec)	HGGA (sec)
20	30	3.82	4.71
	45	4.21	4.73
	60	4.26	5.03
	75	5.62	5.69
30	30	5.33	5.89
	45	12.61	11.77
	60	17.42	13.15
	75	21.07	18.03
40	30	18.97	18.99
	45	28.16	28.15
	60	41.96	41.44
	75	61.32	62.90
50	30	60.71	52.00
	45	73.10	71.32
	60	78.12	77.08
	75	81.06	78.12
60	30	86.48	78.06
	45	89.14	82.78
	60	117.65	81.37
	75	188.52	103.24

n denotes number of customer orders; c is the picker capacity

Considering the computation times of the algorithms, the HGGA performed better than GA in 85 % of the problem instances, that is, 36 out of 40 instances. Out of the 6 instances in which HGGA performed less than GA, only one has customer order above 40. Moreover, the rate of increase of HGGA's computational time over problem size is much less when compared to that of GA. This demonstrates that HGGA is more efficient even over large-scale problems. HGGA's group encoding scheme, unique grouping genetic operators, and constructive heuristics give the algorithm an enhanced computational efficiency. This shows that HGGA has a superior computational efficiency than GA.

7.7 Summary

Optimized order batching is very important for efficient operation of manual order picking systems in distribution warehouses. The chapter presented a hybrid grouping genetic algorithm (HGGA) for the order batching problem. Extensive numerical experiments were used to test the utility of the algorithm. Comparative performance analysis of the algorithm and other benchmark heuristics in the literature showed that HGGA can provide better solutions. In terms of computation times (CPU times), the HGGA computation times were generally shorter when compared to other algorithms.

The proposed HGGA can reduce the length of picker tours significantly. In practice, this demonstrates an effective reduction of the overall picking time, which may translate to cutting down of operational costs and reduction of overtime or workforce. Improved solution quality and computation times also imply that the average lead time for customer orders is also reduced, which ultimately leads to high quality of service. In the long run, this will have a positive impact on the survival of the order picking system and the overall distribution warehouse system.

It will be interesting to carry out further studies the impact of order batching systems on related activities such as article location, picker routing, and warehouse design. It is hoped that such integrated perspectives will greatly improve the overall performance of logistics and warehouse systems.

References

Albareda-Sambola M, Alonso-Ayuso A, Molina E, Simón de Blas C (2009) Variable neighborhood search for order batching in a warehouse. Asia-Pac J Oper Res 26(5):655–683
Bozer YA, Kile JW (2008) Order batching in walk-and-pick order picking systems. Int J Prod Res 46(7):1887–1909
Caron F, Marchet G, Perego A (1998) Routing policies and COI-based storage policies in picker-to-part systems. Int J Prod Res 36(3):713–732
Chen M-C, Wu H-P (2005) An association-based clustering approach to order batching considering customer demand patterns. Omega Int J Manag Sci 33(4):333–343

Chen MC, Huang CL, Chen KY, Wu HP (2005) Aggregation of orders in distribution centers using data mining. Expert Syst Appl 28(3):453–460

Clarke G, Wright JW (1964) Scheduling of vehicles from a central depot to a number of delivery points. Oper Res 12(4):568–581

de Koster R, Roodbergen KJ, van Voorden R (1999a) Reduction of walking time in the distribution center of De Bijenkorf. In: Speranza MG, Stähly P (eds) New trends in distribution logistics. Springer, Berlin, pp 215–234

de Koster R, van der Poort ES, Wolters M (1999b) Efficient order batching methods in warehouses. Int J Prod Res 37(7):1479–1504

de Koster R, Le-Duc T, Roodbergen KJ (2007) Design and control of warehouse order picking: a literature review. Eur J Oper Res 182(2):481–501

Elsayed EA (1981) Algorithms for optimal material handling in automatic warehousing systems. Int J Prod Res 19(5):525–535

Elsayed EA, Unal OI (1989) Order batching algorithms and travel-time estimation for automated storage/retrieval systems. Int J Prod Res 27(7):1097–1114

Gademann N, van de Velde S (2005) Order batching to minimize total travel time in a parallel-aisle warehouse. IIE Trans 37(1):63–75

Gibson DR, Sharp GP (1992) Order batching procedures. Eur J Oper Res 58(1):57–67

Hall RW (1993) Distance approximation for routing manual pickers in a warehouse. IIE Trans 25:77–87

Henn S (2012) Algorithms for on-line order batching in an order picking warehouse. Comput Oper Res 39:2549–2563

Henn S, Wäscher G (2012) Tabu search heuristics for the order batching problem in manual order picking systems. Eur J Oper Res 222:484–494

Henn S, Koch S, Doerner K, Strauss C, Wäscher G (2010) Metaheuristics for the order batching problem in manual order picking systems. Bus Res (BuR) 3(1):82–105

Henn S, Koch S, Wäscher G (2012) Order batching in order picking warehouses: a survey of solution approaches. In: Manzini R (ed) Warehousing in the global supply chain: advanced models, tools and applications for storage systems. Springer, London, pp 105–137

Ho Y-C, Tseng Y-Y (2006) A study on order-batching methods of order-picking in a distribution centre with two cross-aisles. Int J Prod Res 44(17):3391–3417

Ho Y-C, Su T-S, Shi Z-B (2008) Order-batching methods for an order-picking warehouse with two cross aisles. Comput Ind Eng 55(2):321–347

Hsu C-M, Chen K-Y, Chen M-C (2005) Batching orders in warehouses by minimizing travel distance with genetic algorithms. Comput Ind 56(2):169–178

Hwang H, Kim DG (2005) Order-batching heuristics based on cluster analysis in a low-level picker-to-part warehousing system. Int J Prod Res 43(17):3657–3670

Jarvis JM, McDowell ED (1991) Optimal product layout in an order picking warehouse. IIE Trans 23(1):93–102

Kashan AH, Akbari AA, Ostadi B (2015) Grouping evolution strategies: An effective approach for grouping problems. Appl Math Modell 39(9):2703–2720

Petersen CG (1997) An evaluation of order picking routing policies. Int J Oper Prod Manag 17(11):1098–1111

Petersen CG, Aase G (2004) A comparison of picking, storage, and routing policies in manual order picking. Int J Prod Econ 92(1):11–19

Petersen CG, Schmenner RW (1999) An evaluation of routing and volume-based storage policies in an order picking operation. Decis Sci 30(2):481–501

Roodbergen KJ, De Koster R (2001a) Routing methods for warehouses with multiple cross aisles. Int J Prod Res 39(9):1865–1883

Roodbergen KJ, De Koster R (2001b) Routing order-pickers in a warehouse with a middle aisle. Eur J Oper Res 133:32–43

Tompkins JA, White JA, Bozer YA, Tanchoco JMA (2003) Facilities planning, 3rd edn. Wiley, New Jersey

Tsai C-Y, Liou JJM, Huang T-M (2008) Using a multiple-GA method to solve the batch picking problem: considering travel distance and order due time. Int J Prod Res 46(22):6533–6555

Wäscher G (2004) Order picking: a survey of planning problems and methods. In: Dyckhoff H, Lackes R, Reese J (eds) Supply chain management and reverse logistics. Springer, Berlin, pp 323–347

Yu M, de Koster R (2009) The impact of order batching and picking area zoning on order picking system performance. Eur J Oper Res 198(2):480–490

Chapter 8
Fleet Size and Mix Vehicle Routing: A Multi-Criterion Grouping Genetic Algorithm Approach

8.1 Introduction

The need for efficient and effective transportation services is ever increasing all over the globe. Transportation costs account for at least 20 % of the total cost of a product (Hoff et al. 2010). On the other hand, the transportation sector in every economy contributes a significant portion to the gross domestic product of every economy (Engevall et al. 2004; Hoff et al. 2010). In the same vein, the sector employs several millions of people worldwide. Since the 1980s, the world's maritime fleet has grown by more than 25 %, and productivity has increased by 12.5 % in the same period (Hoff et al. 2010; Christiansen et al. 2007). The sector continues to grow as the demand for its services continues to grow.

There is an ever-increasing global competition in the logistics and transportation sector. This constantly aggravates the demand for efficiency, quality of service, timeliness, agility, and cost-effectiveness in the sector. Pressure from various stakeholders is also a concern for transportation and logistics services. The society has become aware of the climate and environmental issues. Customers and other stakeholders continue to call for green vehicle routing and efficient logistics operations planning that contribute to reduced environmental damage (Erdogan and Miller-Hooks 2012). As such, decision makers in transportation and logistics are left with no option except to optimize their operations.

As decision makers strive to optimize the planning and utilization of vehicles to serve networks of hundreds of customers at various locations, decision criteria such as transportation cost and quality of routing schedule are important for cost-effective decisions. While it is crucial to utilize cost, it is also important to consider balancing the workload assignment among the vehicles or drivers. In fleet size and mix vehicle routing, a common problem in transportation, distribution, and logistics, a fleet of delivery vehicles should serve customers with commodities from a common depot. Central to these operations is the issue of planning for and

© Springer International Publishing Switzerland 2017
M. Mutingi and C. Mbohwa, *Grouping Genetic Algorithms*,
Studies in Computational Intelligence 666,
DOI 10.1007/978-3-319-44394-2_8

scheduling of a fleet of vehicles, which can consume quite a great deal of time and effort, especially in the presence of multiple criteria.

The problem can be viewed as an extension of the classical vehicle routing problem: a variant of the capacitated vehicle routing problem, concerned with how to determine the number of vehicles of each type to be used, for a given mix of vehicle types of different capacity and costs. An important assumption is that the fleet is heterogeneous, and the available number of vehicles for each type is unlimited. As such, the concern is to make a decision on fleet composition as well as vehicle routing.

The purpose of this chapter was to develop a multiple criteria approach to fleet size and mix vehicle routing. By the end of this chapter, the reader is expected to learn the following:

1. To understand the fleet size and mix vehicle routing problem in the presence of multiple criteria;
2. To conceptualize the problem as a grouping problem with a typical grouping structure;
3. To implement a computational approach to the multi-criterion grouping problem; and
4. To carry out computational experiments and interpret the managerial implications of the results obtained.

In the presence of a myriad of constraints, multiple possible combinations of vehicle types and routing patterns, and multiple criteria, obtaining a high-quality plan and a routing schedule is a complex decision problem.

The rest of this chapter is organized as follows: The next section presents a description of the fleet size and mix vehicle routing problem. Section 8.3 provides extant related work in the literature. Section 8.4 describes the proposed multi-criterion grouping genetic algorithm. Section 8.5 presents computational tests and discussions. Section 8.6 provides a summary of the chapter.

8.2 Fleet Size and Mix Vehicle Routing Problem Description

The fleet size and mix vehicle routing problem (FSMVRP) are described as follows: A set of n customers, $\{1, 2,\ldots, n\}$, at different locations, are to be served by a fleet of T vehicle types available at the depot, represented by 0. Assume that the number of vehicles of each type is unlimited. Thus, one of the decisions is to determine the number of vehicles of each type t with capacity Q_t, a fixed cost f_t, and a variable cost per unit distance v_t. Also, assume that between any two vehicle types a and b, $f_a < f_b$, if $Q_a < Q_b$. There are usually two cost structures: (i) different fixed costs with uniform variable costs and (ii) different variable costs with no fixed costs (Liu et al. 2009). Therefore, with fixed costs and uniform variable costs,

$$v_t = 1; \quad f_t > 0 \quad \forall t \tag{8.1}$$

and with variable costs and no fixed costs,

$$v_t > 0; \quad f_t = 0 \quad \forall t \tag{8.2}$$

Each customer node $i > 0$ has a nonnegative demand d_i. Denote the traveling distance between locations i and j by nonnegative τ_{ij}, which is symmetric and satisfies the triangle inequality, $\tau_{ij} = \tau_{ji}$ and $\tau_{ij} + \tau_{jk} \geq \tau_{ik}$. It follows that the total variable cost of traveling from i to location j is $v_t \tau_{ij}$. The aim of the FSMVRP is to determine the vehicle fleet composition and the route of each vehicle, so that the total cost of delivering goods to all customers is minimized and the workload is balanced, subject to conditions: (i) Each route starts and ends at the depot; (ii) each customer is visited exactly once; (iii) customer demands are satisfied; and (iv) vehicle capacity is not violated.

Apart from the basic vehicle routing features, it can be seen from the above description that the FSMVRP possesses very unique complicating features, including the following:

1. The problem has a unique grouping structure with order-dependent restrictions and repeatable group identities;
2. The presence of multiple decision criteria and multiple vehicle capacity constraints;
3. The problem is a combination of two problems, namely vehicle planning and vehicle routing problems.

These features pose serious challenges to the development of computational approaches to the problem. The following section presents the related work in the literature.

8.3 Related Work

Vehicle routing has been the central issue among researchers. This section provides a background to vehicle routing and presents related approaches to solving the fleet size and mix vehicle routing problem.

8.3.1 Vehicle Routing: A Background

The FSMVRP can be traced back to the classical vehicle routing problem (VRP), first studied by Dantzig and Ramser (1959). Basically, the purpose of the VRP is to minimize transportation costs, number of vehicles used, and the customer waiting times (Taillard 1999; Tarantilis et al. 2004; Toth and Vigo 2012; Salhi and Rand 1993;

Lima et al. 2016; Koç et al. 2015, 2016). Since its inception of the VRP, several challenging extensions to the problem have emerged over the years. The capacitated vehicle routing problem (CVRP) is an extended version of the VRP that focuses on optimizing the dispatch of customer goods using a fleet of capacitated homogenous vehicles (Lima et al. 2016; Ai and Kachitvichyanukul 2009).

Further to the above, vehicle routing problem with time window constraints (VRPTW) is a common problem in logistics and distribution. VRPTW seeks to avoid service delivery delays by minimizing arrivals after the latest time window, which in turn maximizes the service quality (Wang et al. 2015; Braysey and Gendreau 2005). The heterogeneous fixed fleet vehicle routing problem (HFFVRP) is a variant of CVRP where the number of available vehicles is fixed, the vehicles are of different types, and the decision is to determine the best way to utilize the existing fleet of vehicles (Avci and Topaloglu 2016; Wang et al. 2015). The fleet size is fixed or bounded by the maximum number of each vehicle type. Contrary to the HFFVRP, the FSMVRP is a variant of the CVRP, where the fleet size and its composition are to be determined (Wassan and Osman 2002; Desrochers and Verhoog 1991; Renaud and Boctor 2002; Choi and Tcha 2007). Furthermore, the FSMVRP with time windows is also an extension of the FSMVRP (Braysey and Gendreau 2005). Each of the above variants of the VRP can also have single or multiple depots.

8.3.2   Approaches to Fleet Size and Mix Vehicle Routing

Due to the ever-increasing complexity of the FSMVRP, the application of exact mathematical programming methods is not a viable option, especially with large-scale problems. The FSMVRP is a NP-hard problem. Several classical heuristics have been designed to address the FSMVRP, such the saving algorithm (Clarke and Wright 1964), the sweep algorithm (Gillett and Miller 1974) and the generalized assignment (Fisher and Jaikumar 1981) and the matching-based saving algorithms (Desrochers and Verhoog 1991; Osman and Salhi 1996). However, their application is limited, especially over large-scale problems.

Mathematical programming-based heuristics have been proposed by several researchers in the literature, including Taillard (1999), Renaud and Boctor (2002), Yaman (2006), and Choi and Tcha (2007). For instance, Choi and Tcha (2007) developed a set covering formulation and solved its linear relaxation by column generation to obtain the bounds and obtained some of the best results of the benchmark problems in the literature. However, decision makers often desire efficient solution approaches that can produce near-optimal solutions within short computation times.

As problem sizes continue to increase, a number of researchers have utilized metaheuristic approaches to obtain optimal or near-optimal solutions. Such methods include tabu search (Wassan and Osman 2002), memetic algorithm (Lima et al. 2004), particle swarm optimization (Moghadam and Seyedhosseini 2010; Ai and Kachitvichyanukul 2009), genetic algorithm (Ochi et al. 1998), and other evolutionary algorithms (Wang et al. 2008). Further developments along these lines are more promising.

The next section goes a step further to addressing the multiple criteria FSMVRP using an improved multi-criterion grouping genetic algorithm (GGA) with a unique grouping representation scheme and genetic operators (selection, crossover, mutation, and inversion).

8.4   Multi-Criterion Grouping Genetic Algorithm Approach

The multi-criterion GGA and its elements are described, including chromosome coding, initialization, and genetic grouping operators (selection, crossover, mutation, and inversion).

8.4.1   GGA Encoding

The type of the coding scheme implemented in GGA has a strong influence on the efficiency of the algorithm. Most researchers have used depot(s) as trip delimiters in their GA encoding schemes (Ochi et al. 1998; Lima et al. 2004; Prins 2004). In GGA encoding, a chromosome $k = [1, 2, 3,\ldots, n]$ is partitioned into groups of customer orders so that the delivery cost incurred is minimized, and the total load for each group (i.e., each trip) does not exceed the capacity of the vehicle type assigned to that trip. This can be expressed in the form of an acyclic graph G (X) with vertex $V(G) = \{i \mid 0 \leq i \leq n\}$, where $E(G)$ represents a set of directed arcs on $G(X)$, $(i, j) \in E(G)$ if $\sum_{m=i+1}^{j} d_m \leq Q_t$ represents a feasible trip assigned to vehicle type t that departs from the depot (node $i = 0$), visits nodes $i + 1, i + 2,\ldots,$ $j - 1$, and j, consecutively, and returns to the depot. The total load for a trip (i, j) is given by $\sum_{m=i+1}^{j} d_m$. The objective is to select a vehicle type t with the least cost and capacity not less than the trip load. Then, with fixed cost, the trip cost c_{ij} is equivalent to fixed cost plus variable costs (Liu et al. 2009):

$$c_{ij} = f_t + \tau_{0,i+1} + \sum_{h=i+1}^{j-1} \tau_{h,h+1} + \tau_{j,0} \qquad (8.3)$$

and with no fixed cost,

$$c_{ij} = \left(\tau_{0,i+1} + \sum_{h=i+1}^{j-1} \tau_{h,h+1} + \tau_{j,0} \right) \cdot v_t \qquad (8.4)$$

It can be seen that the shortest path on $G(X)$ defines the optimal partition of the sequence. An optimal decoding can be found in $O(n^2\log(n))$ time, and a faster algorithm can be designed to solve this shortest path problem in $O(n^2)$ or even $O(n)$ time (Prins 2004).

Figure 8.1 considers a distribution center with $T = 2$ unlimited vehicle types to serve $n = 6$ retailers. The respective capacities and fixed costs of the vehicle types t_1 and t_2 are $Q_1 = 500, Q_2 = 550,$ and $f_1 = 300$ and $f_2 = 400$. The numbers on arc(i,j) and node j represent τ_{ij} and d_i, respectively. The proposed GGA encoding uses a group structure comprising three parts, as indicated in Fig. 8.2. The items part shows the customers to be grouped, which in turn corresponds to a trip; the group part encodes the vehicle type assigned to each trip. Finally, the frontier part codes the position of the last node of each trip. Note that since groups represent vehicle type, some groups can appear more than once in the group structure. This is a unique feature of the grouping problem.

It is essential to realize that the group represents the vehicle type, which implies that the group part of the chromosome can have repeated group identities, as in this example. In addition, the coding is order dependent; for example, vehicle 2 (group 2) is supposed to visit three customers 3, 5, and 6 in that sequence.

According to the coding, vehicle type 1, with $f_1 = 300$ and $v_t = 1$, is assigned trip $\{0{-}1{-}2{-}0\}$. Therefore, from (8.3), the trip cost is given by $300 + 240 + 45 + 280 = 865$. The other trips, $\{0{-}4{-}0\}$ and $\{0{-}3{-}5{-}6{-}0\}$, are evaluated in the same way, as summarized in Table 8.1. Let the total cost be denoted by F_1. Then, the total cost for all the trips is $F_1 = 2492$.

To balance the workload assignment among the workers (or vehicles), the variation of each individual workload from the average workload should be minimized. Assume that the average workload is w and each individual workload is denoted by ω_j; then, the total cost due to workload variation is:

$$F_2 = \sum_{j=1}^{g} c \cdot |\omega_j - w| \tag{8.5}$$

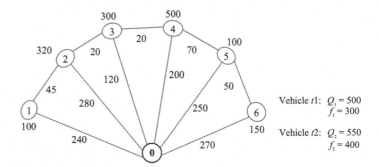

Fig. 8.1 Typical data for chromosome representation

Fig. 8.2 Chromosome
coding for the FSMVRP

Items:	1	2	4	7	3	5	6
Group:		1		1		2	
Frontier:		2		3		6	

Table 8.1 AGGA coding
solution example

Trip	Vehicle type	Cost
0–1–2–0	1	865
0–4–0	1	700
0–3–5–6–0	2	930
Total cost		2492

where the c is a pseudo-cost for workload variation. It follows that the overall objective function F can be expressed as a weighted sum of the trip costs (F_1) plus workload variation costs (F_2), as follows:

$$F = \lambda_1 \cdot F_1 + \lambda_2 \cdot F_2 \qquad (8.6)$$

where λ_1 and λ_2 are the weights of the trip costs and workload variation costs, respectively.

8.4.2 Initialization

In this GGA implementation, the initial population of p chromosomes is generated using (i) the savings and sweep heuristics (Gillett and Miller 1974) and (ii) random generation. As such, the savings algorithm is applied using one vehicle type at a time. These initial solutions are then concatenated into chromosomes. More initial chromosomes are generated using the following algorithm:

Algorithm 1 Population initialization

1. Input: population size p; directed graph $G(X)$
2. Savings $(G(X))$;
3. Sweep $(G(X))$;
4. **Repeat**
5.   Assign a node to each vehicle t, $(t = 1,2,\ldots, m)$;
6.   Randomly assign the remaining nodes;
7.   Encode the chromosome and add to initial population;
8.   p++;
9. **Until** (population size $= p$).
9. **Return** population;

The GGA seeks to minimize some cost function by mapping the cost function onto an evaluation function before selecting the best performing candidate solutions for crossover.

8.4.3 Selection

In the selection procedure, the remainder stochastic sampling without replacement was applied; each chromosome k is selected for mating according to its expected count e_k which is calculated according to the expression:

$$e_k = a \cdot f_k \bigg/ \sum_{k=1}^{P} \left(\frac{f_k}{p}\right) \qquad (8.7)$$

where f_k is the score function of the kth chromosome and $a \in [0,1]$ is an adjustment parameter. Each chromosome receives copies equal to the integer part of e_k and additional copies through a success probability equivalent to the fractional part of e_k. The copies are added into the mating pool for the group crossover operation.

8.4.4 Crossover

The group crossover operator is an evolutionary mechanism through which selected chromosomes mate to produce a population of new offspring, called selection pool. Crossover enables GGA to explore unvisited regions in the solution space, a process called exploration. A two-point group crossover is applied (at a probability p^c) repeatedly, using different parent chromosomes until the desired pool size, *poolsize*, is obtained. The crossover operator is listed in Algorithm 2:

Algorithm 2 Selection operator
1. **Repeat**
2.   Select two parent chromosomes, P_1 and P_2, $P_1 \neq P_2$
3.   Select crossing sections for P_1 and P_2
4.   Cross P_1 and P_2 to obtain offspring O_1 and O_2
5.   Eliminate *doubles*, avoiding crossed items
6.   Insert *misses*, beginning from where doubles were eliminated
7.   Repair the offspring, if necessary
8. **Until** (selection *poolsize* is achieved)
9. **Return**

Figure 8.3 provides an example of the crossover mechanism based on two randomly selected parent chromosomes, namely $P_1 = [1\ 2\ 3]$ with groups of items $\{1,2\}$, $\{4,7\}$, and $\{3,5,6\}$, and $P_2 = [4\ 5\ 6]$ with groups of items $\{1,3\}$, $\{2,5,7\}$, and $\{4,6\}$.

In the crossover process, some customers may appear in more than one trip, while others may be missing. Thus, the offspring is repaired by eliminating

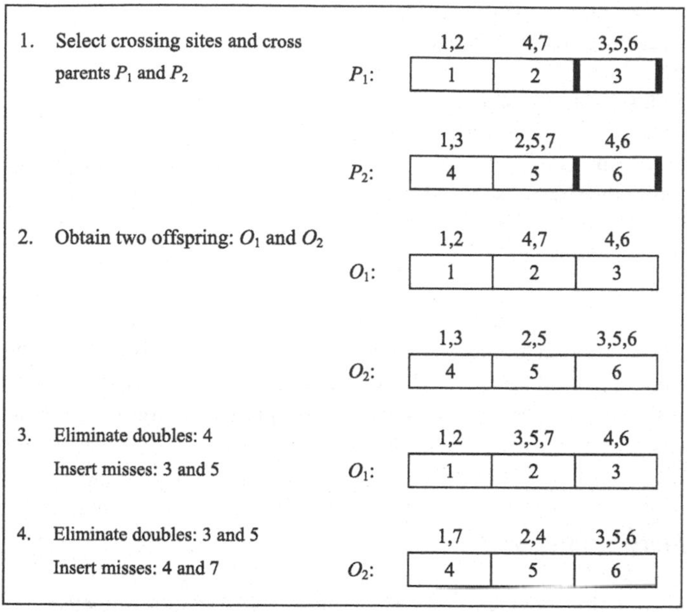

1.	Select crossing sites and cross parents P_1 and P_2	P_1:	1,2	4,7	3,5,6
			1	2	3

		P_2:	1,3	2,5,7	4,6
			4	5	6

2.	Obtain two offspring: O_1 and O_2	O_1:	1,2	4,7	4,6
			1	2	3

		O_2:	1,3	2,5	3,5,6
			4	5	6

3.	Eliminate doubles: 4 Insert misses: 3 and 5	O_1:	1,2	3,5,7	4,6
			1	2	3

4.	Eliminate doubles: 3 and 5 Insert misses: 4 and 7	O_2:	1,7	2,4	3,5,6
			4	5	6

Fig. 8.3 Group crossover operation

duplicated customers (doubles) and inserting missing ones (misses), preferably the ones with the least loading. After crossover, a new population, *newpop*, is created by combining the best performing offspring and the current population *oldpop*.

8.4.5 Mutation

In this GGA implementation, mutation is applied to every new chromosome using two mutation operators, that is, the merge mutation and the split mutation.

8.4.5.1 Split Mutation

Split mutation selects a group and splits it into two groups, at a probability q_j ($j = 1...m$), where q_j is directly proportional to the size of the group. Large groups are expected to reduce, while small groups should increase in size. This promotes mutation of groups to a balanced state. Each item i from the split group j' is assigned to any other group j at a probability p_{ij}, a function of the similarity coefficient γ_{ij} between item i and group j. The algorithm for the split mutation is as follows:

Algorithm 3 Split mutation procedure
1. Input: population size p, population
2. **For** count = 1 to p
3.   **If** {prob $(p^m$ = true)} **Then**
4.   Select a group j' at a probability q_j
5.   Re-assign every item i from j' to any other group j at probability p_{ij}
6.   Repair the chromosome, if necessary
7.   **End**
6. **End**
7. **Return**

An example of split mutation is presented in Fig. 8.4, based on chromosome [1 2 3] with groups {1,3}, {5,6}, and {2,4,7,8}. In this application, the split mutation is designed to work together with the merge mutation, which is described next.

8.4.5.2 Merge Mutation

The merge mutation consists in concatenating two groups selected based on an adaptive probability. As such, the probability of choosing the group is a function of size of the group; the higher the group size, the lower the probability of being selected for the merge mutation. Thus, the probability q_j for selecting group j for mutation is a function of the size of the group. Similar to the split mutation mechanism, increasing the chances of merging small groups promotes balanced grouping, which is essential in most real-world problems, such as in home healthcare staff scheduling (Mutingi and Mbohwa 2016) and task assignment (Muting and Mbohwa 2014), where fairness in workload assignment is crucial for high-quality schedules. The merge mutation operator can be summarized as in the following algorithm:

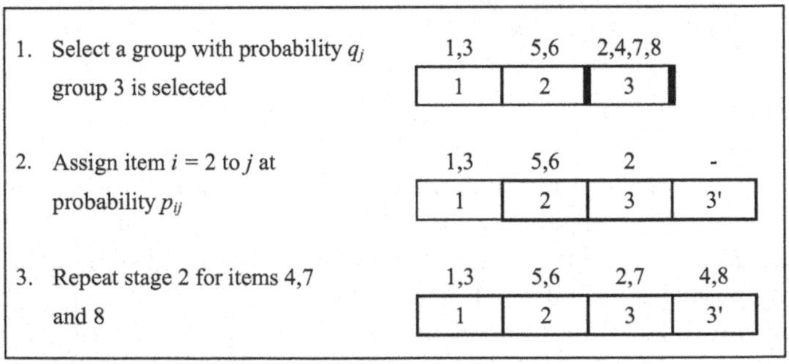

		1,3	5,6	2,4,7,8	
1.	Select a group with probability q_j group 3 is selected	1	2	3	

		1,3	5,6	2	-
2.	Assign item i = 2 to j at probability p_{ij}	1	2	3	3'

		1,3	5,6	2,7	4,8
3.	Repeat stage 2 for items 4,7 and 8	1	2	3	3'

Fig. 8.4 Split mutation for the FSMVRP

Algorithm 4 Merge mutation
1. Input: population size p, population
2. **For** count = 1 to p
3.   **If** {prob(p^m) = true} **Then**
4.   Select two groups, $j1$ and $j2$, with probabilities q_{j_1} and q_{j_2}, respectively
5.   Merge the two groups, $j1$ and $j2$ into one group
6.   Re-assign items at probability p_{ij} if any hard constraint is violated
7.   **End**
6. **End**
4. **Return**

Figure 8.5 provides an example of mutation by merging, using chromosome [{1,3}{6,9,2}{4}{8,5,7}]. The resulting chromosome is [{6,9,2}{1,4,3}{8,1,7}], with evenly distributed group sizes. If a grouping constraint is violated in the mutation process, the chromosome is restructured by reassigning each unassigned item i to group j at a probability p_{ij}.

8.4.6   Inversion

Inversion is applied at a very low probability p^i on randomly selected chromosomes. The operator enhances the crossover operator by improving the chances of involving shifted groups. Groups that are very close together in the sequence have higher chances of being shifted together. In this application, the two-point inversion is implemented.

The two-point inversion overturns items within two randomly selected inversion sites. If any hard constraints are violated by the resulting chromosome, a repair mechanism is utilized. The algorithm for the swap mechanism is presented as follows:

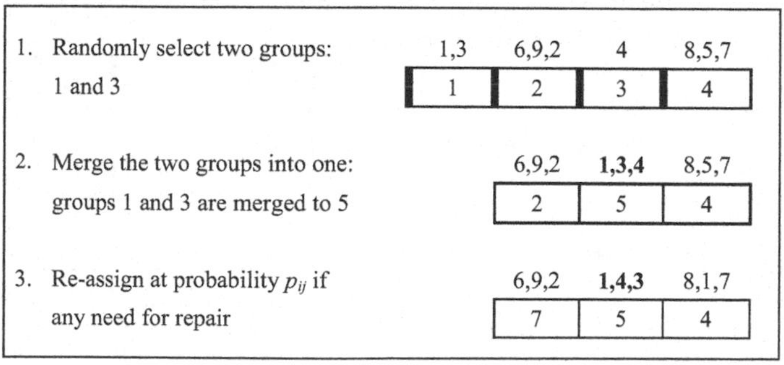

1. Randomly select two groups:	1,3	6,9,2	4	8,5,7
1 and 3	1	2	3	4
2. Merge the two groups into one:		6,9,2	**1,3,4**	8,5,7
groups 1 and 3 are merged to 5		2	5	4
3. Re-assign at probability p_{ij} if		6,9,2	**1,4,3**	8,1,7
any need for repair		7	5	4

Fig. 8.5 Merge mutation for the FSMVRP

Algorithm 5 Swap inversion
1. Input: population, p; inversion probability p^i
2. **For** $i = 1$ to p **do**
3.   **If** {prob (p^i)} **Then**
4.   Randomly select two inversion points;
5.   Overturn the selected groups;
6.   Repair, if need be;
7.   **End**
8. **End**
9. **Return**

Figure 8.6 illustrates the two-point inversion procedure, based on an example of the chromosome [{1,3}{4,7}{2,6}{8,5}]. Groups 2 and 3 lying between the two inversion sites are overturned. As a result, chromosome [{1,3}{4,7}{2,6}{8,5}] is mutated to [{1,3}{2,6}{4,7}{8,5}]. This implies that chromosome [1 2 3 4] is transformed to [1 3 2 4] by writing the groups within the inversion sites in the reverse order. If any infeasible solution is generated, a repair mechanism is applied.

8.4.7 Diversification

As iterations proceed, the population may prematurely converge to a particular solution which is far from optimal. To prevent premature convergence, population diversity should be checked at every iteration. First, define an entropic measure H_i for each customer (or node i);

$$H_i = \frac{1}{\log(m)} \sum_{j=1}^{m} (n_{ij}/p) \cdot \log (n_{ij}/p) \tag{8.8}$$

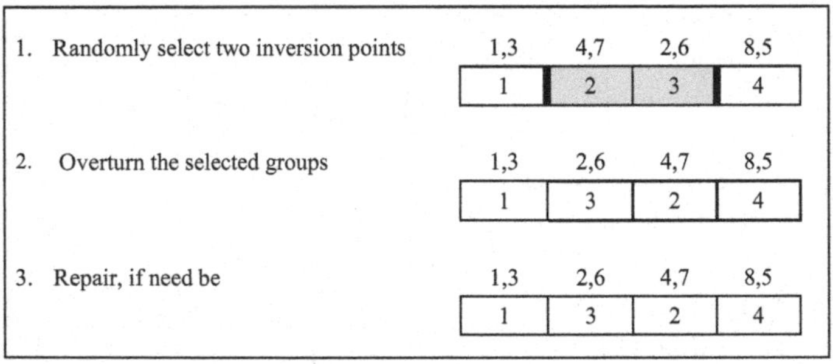

1. Randomly select two inversion points	1,3	4,7	2,6	8,5
	1	2	3	4

2. Overturn the selected groups	1,3	2,6	4,7	8,5
	1	3	2	4

3. Repair, if need be	1,3	2,6	4,7	8,5
	1	3	2	4

Fig. 8.6 Two-point inversion operator for the FSMVRP

where n_{ij} is the number of chromosomes in which customer i is assigned position j in the current population; m is the number of customers. Then, diversity H is determined thus:

$$H = \sum_{i=1}^{m} H_i/m \tag{8.9}$$

Therefore, inversion is applied to improve diversity to a desired value H_{min}. The best candidates from diversified and undiversified populations are always preserved. In the case where diversity H falls below acceptable value, full inversion is applied till diversity is acceptable. An example of full inversion is presented based on chromosome $[1 \quad 2 \quad 3 \quad 4 \quad 5]$,

Before full inversion, $[1 \quad 2 \quad 3 \quad 4 \quad 5]$
After full inversion, $[5 \quad 4 \quad 3 \quad 2 \quad 1]$

It is desirable to let the acceptable H_{min} decay over iteration count, so that the population can be allowed to converge in the long run. Thus, a decay function is applied. The next section presents a summary of the GGA computational implementation.

8.4.8 GGA Computational Implementation

The overall GGA incorporates the operators described in previous sections, using carefully chosen genetic probabilities, based on our experiments: crossover (0.42), mutation (0.06), and inversion (0.04). The algorithm begins by accepting basic input from the user. The overall pseudo-code for the GGA procedure is presented in Algorithm 6:

Algorithm 6 GGA implementation
1. **Input**: initial data input: GGA parameters;
2. Initialize population, *oldpop*: create chromosomes;
3. **Repeat**
4.   Evaluation ()
5.   Crossover ();
6.   Mutation ();
7.   Inversion ();
8. **Until** (*gen* ≥*maxGen*);
9. **End**
10. **Return**: Solutions.

Here, the selection procedure is assumed to be included in the crossover mechanism. Other genetic mechanisms such as replacement strategy and diversification are not explicitly presented in the pseudo-code.

8.5 Computational Tests and Discussions

Computational experiments, analysis, and pertinent discussions are presented in this section.

8.5.1 Computational Experiments

The proposed GGA was implemented in Java and executed on a Pentium 4 at 3 GHz based on 12 benchmark problems in Golden et al. (1984). Table 8.2 specifies the costs, vehicle capacity Q_t, ($t = 1, 2,..., 6$), fixed cost f_t, and variable cost v_t. Using the notation in (Taillard 1999), problem sets 3–6 have 20 customers, 13–16 have 50, 17–18 have 75, and 19–20 have 100. Computational results were compared with those from best performing heuristics in the literature, including tabu search (Gendreau et al. 1999; Wassan and Osman 2002; Brandao 2008), column generation-based heuristics (Choi and Tcha 2008), and evolutionary algorithm (Ochi et al. 1998).

8.5.2 Computational Results and Discussions

Table 8.3 presents the computational results of the FSMVRP with fixed costs. A count of the best known solutions obtained by the heuristics is provided. Out of the 12 benchmark problems, our GGA approach produced 11 best known solutions, compared to only 6 found in Gendreau et al. (1999), 6 in Wassan and Osman (2002), 5 in Lima et al. (2004), 8 in Choi and Tcha (2007), 9 in Brandao (2008), and 10 in Liu et al. (2009).

Table 8.4 presents the percentage deviation of each solution from the best known and the computation times for each problem. All algorithms showed remarkable accuracy, with average percent deviation less than 1 %. Our GGA performed competitively in terms of percentage deviation and computation times. The results demonstrate the utility of the GGA developed in this research.

Table 8.2 Specifications for the benchmark problems

No.	Q_1	f_1	v_1	Q_2	f_2	v_2	Q_3	f_3	v_3	Q_4	f_4	v_4	Q_5	f_5	v_5	Q_6	f_6	v_6
3	20	20	1.0	30	35	1.0	40	50	1.0	70	120	1.0	120	225	1.0			
4	60	1000	1.0	80	1500	1.0	150	3000	1.0									
5	20	20	1.0	30	35	1.0	40	50	1.0	70	120	1.0	120	225	1.0			
6	60	1000	1.0	80	1500	1.0	150	3000	1.0									
13	20	20	1.0	30	35	1.1	40	50	1.2	70	120	1.7	120	225	2.5	200	400	3.2
14	120	100	1.0	160	1500	1.1	300	3500	1.4									
15	50	100	1.0	100	250	1.6	160	450	2.0									
16	40	100	1.0	80	200	1.6	140	400	2.1									
17	50	25	1.0	120	80	1.2	200	150	1.5	350	320	1.8	250	400	2.9	400	800	3.2
18	20	10	1.0	50	35	1.3	100	100	1.9	150	180	2.4						
19	100	500	1.0	200	1200	1.4	300	2100	1.7									
20	60	100	1.0	140	300	1.7	200	500	2.0									

Table 8.3 Computational results for the benchmark problems

No.	Best known	Gendreau et al. (1999)	Wassan and Osman (2002)	Lima et al. (2004)	Choi and Tcha (2007)	Brandao (2008)	Liu et al. (2009)	GGA
3	961.03	961.03	961.03	961.03	961.03	961.03	961.03	961.03
4	6437.33	6437.33	6437.33	6437.33	6437.33	6437.33	6437.33	6437.33
5	1007.05	1007.05	1007.05	1007.05	1007.05	1007.05	1007.05	1007.05
6	6516.47	6516.47	6516.47	6516.47	6516.47	6516.47	6516.47	6516.47
13	2406.36	2408.41	2422.10	2408.60	2406.36	2406.36	2406.36	2406.36
14	9119.03	9119.03	9119.86	9119.03	9119.03	9119.03	9119.03	9119.03
15	2586.37	2586.37	2586.37	2586.88	2586.37	2586.37	2586.37	2586.37
16	2720.43	2741.5	2730.08	2721.76	2720.43	2728.14	2724.22	2720.43
17	1734.53	1749.5	1755.1	1758.53	1758.53	1734.53	1734.53	1734.53
18	2369.65	2381.43	2385.52	2396.47	2371.49	2369.65	2369.65	2369.65
19	8659.74	8675.16	8659.74	8691	8664.29	8661.81	8662.95	8661.81
20	4038.46	4086.76	4061.64	4093.29	4039.49	4042.59	4038.46	4038.46
Bests	12	6	6	5	8	9	10	11

8.6 Summary

The FSMVRP is concerned with the determination the fleet size and the composition or mix of heterogeneous vehicles. It is assumed that the number of vehicles of each type is unlimited. This chapter presents a multi-criterion GGA for solving the multi-criterion FSMVRP with fixed and variable costs. Computational experiments were conducted based on benchmark problems in the literature. Comparative analysis showed that the GGA approach obtained best known solutions. Moreover, the GGA approach performed competitively in terms of computation time. In terms of the average cost, GGA also demonstrated competitive performance.

The work presented offers useful research contributions for the logistics and transportation industry. The proposed GGA approach uses unique grouping genetic operators. When compared to related approaches in the literature, the algorithm demonstrates its competitive performance. Further research directions should include the design of more efficient and flexible algorithms for solving the FSMVRP problems in which the customer demand and/or time window is uncertain or fuzzy. It may also be fruitful to research further on FSMVRP with fuzzy time windows. The approach can also be extended to similar problems such as home healthcare staff schedule and heterogeneous fixed fleet vehicle routing problem.

Table 8.4 Percent deviations and CPU times for FSMVRP algorithms

No.	Best known	Gendreau et al. (1999)		Lima et al. (2004)		Choi and Tcha (2007)		Liu et al. (2009)		GGA (2012)	
		Deviation (%)	Time (s)	Deviation (%)	Time (s)	Deviation (%)	Time (s)	Deviation (%)	Time (s)	Deviation (%)	Time (s)
3	961.03	0.000	164	0.000	89	0.000	0	0.000	0	0.000	0
4	6437.33	0.000	253	0.000	85	0.000	1	0.000	0	0.000	1
5	1007.05	0.000	164	0.000	85	0.000	1	0.000	2	0.000	2
6	6516.47	0.000	309	0.000	85	0.000	0	0.000	0	0.000	0
13	2406.36	0.085	724	0.093	559	0.000	10	0.000	91	0.000	89
14	9119.03	0.000	1033	0.000	669	0.000	51	0.000	42	0.000	65
15	2586.37	0.000	901	0.020	554	0.000	10	0.000	48	0.000	55
16	2720.43	0.775	815	0.049	507	0.000	11	0.139	107	0.000	10
17	1734.53	0.863	1022	1.384	1517	1.384	207	0.000	109	0.000	113
18	2369.65	0.497	691	1.132	1613	0.078	70	0.000	197	0.000	211
19	8659.74	0.178	1687	0.361	2900	0.053	1179	0.037	778	0.024	804
20	4038.46	1.196	1421	1.358	2383	0.026	264	0.000	1004	0.000	1047
Averages		0.3012	776.9	0.3171	887.4	0.1462	172.4	0.0176	209.0	0.0024	209.7

References

Ai J, Kachitvichyanukul V (2009) Particle swarm optimization and two solution representations for solving the capacitated vehicle routing problem. Comput Ind Eng 56:380–387

Avci M, Topaloglu S (2016) A hybrid metaheuristic algorithm for heterogeneous vehicle routing problem with simultaneous pickup and delivery. Expert Syst Appl 53:160–171

Brandao J (2008) A deterministic tabu search algorithm for the fleet size and mix vehicle routing problem. Eur J Oper Res 195(3):716–728

Braysey O, Gendreau M (2005) Vehicle routing problems with time windows, Part I: Route construction and local search algorithms. Transp Sci 39(1):104–118

Choi E, Tcha DW (2007) A column generation approach to the heterogeneous fleet vehicle routing problem. Comput Oper Res 34:2080–2095

Christiansen M, Fagerholt K, Ronen D, Nygreen B (2007) Maritime transportation. In: Barnhart C, Laporte G (eds) Handbook in operations research and management science. Elsevier, Amsterdam, pp 189–284

Clarke G, Wright JW (1964) Scheduling of vehicles from a central depot to a number of delivery points. Oper Res 12:568–581

Dantzig GB, Ramser JH (1959) The truck dispatching problem. Manage Sci 6:80–91

Desrochers M, Verhoog TW (1991) A new heuristic for the fleet size and mix vehicle routing problem. Comput Oper Res 18:263–274

Engevall S, Gothe-Lundgren M, Varbrand P (2004) The heterogeneous vehicle routing game. Transp Sci 38:71–85

Erdogan S, Miller-Hooks E (2012) A green vehicle routing problem. Transp Res Part E 48: 100–114

Fisher M, Jaikumar R (1981) A generalized assignment heuristic for vehicle routing. Networks 11:109–124

Gendreau M, Laporte G, Musaraganyi C, Taillard ED (1999) A tabu search heuristic for the heterogeneous fleet vehicle routing problem. Comput Oper Res 26:1153–1173

Gillett B, Miller L (1974) A heuristic for the vehicle dispatching problem. Oper Res 22:340–349

Golden B, Assad A, Levy L, Gheysens F (1984) The fleet size and mix vehicle routing problem. Comput Oper Res 11:49–66

Hoff A, Andersson H, Christiansen M, Hasle G, Løkketangen A (2010) Industrial aspects and literature survey: fleet composition and routing. Comput Oper Res 37:2041–2061

Koç Ç, Bektaş T, Jabali O, Laporte G (2015) A hybrid evolutionary algorithm for heterogeneous fleet vehicle routing problems with time windows. Comput Oper Res 64:11–27

Koç Ç, Bektaş T, Jabali O, Laporte G (2016) The fleet size and mix location-routing problem with time windows: formulations and a heuristic algorithm. Eur J Oper Res 248(1):33–51

Lima CMRR, Goldbarg MC, Goldbarg EFG (2004) A memetic algorithm for the heterogeneous fleet vehicle routing problem. Electron Notes Discrete Math 18:171–176

Lima FMS, Pereira DSD, Conceição SV, Nunes NTR (2016) A mixed load capacitated rural school bus routing problem with heterogeneous fleet: algorithms for the Brazilian context. Expert Syst Appl 56:320–334. Available online 17 March 2016

Liu S, Huang W, Ma H (2009) An effective genetic algorithm for the fleet size and mix vehicle routing problems. Transp Res Part E 45:434–445

Moghadam BF, Seyedhosseini SM (2010) A particle swarm approach to solve vehicle routing problem with uncertain demand: a drug distribution case study. Int J Ind Eng Comput 1:55–66

Mutingi M, Mbohwa C (2012b) Enhanced group genetic algorithm for the heterogeneous fixed fleet vehicle routing problem. IEEE international conference on industrial engineering and engineering management, Hong Kong, 10–13 Dec 2012, pp 207–2011

Mutingi M, Mbohwa C (2014) A Fuzzy-based particle swarm optimization approach for task assignment in home healthcare. South African J Ind Eng 25(3):84–95

Mutingi M, Mbohwa C (2016) Fuzzy grouping genetic algorithm for homecare staff scheduling. In: Mutingi M and Mbohwa C (ed) Healthcare Staff Scheduling: Emerging Fuzzy Optimization Approaches, 1st edn. CRC Press, Taylor & Francis, New York, 119–136

Ochi LS, Vianna DS, Drummond LM, Victor AO (1998) A parallel evolutionary algorithm for the vehicle routing problem with heterogeneous fleet. Future Gener Comput Syst 14:285–292

Osman S, Salhi S (1996) Local search strategies for the vehicle fleet mix problem. In: Rayward-Smith VJ, Osman IH, Reeves CR, Smith GD (eds) Modern heuristic search methods. Wiley, New York, pp 131–153

Prins C (2004) A simple and effective evolutionary algorithm for the vehicle routing problem. Comput Oper Res 31:1985–2002

Renaud J, Boctor FF (2002) A sweep-based algorithm for the fleet size and mix vehicle routing problem. Eur J Oper Res 140:618–628

Salhi S, Rand GK (1993) Incorporating vehicle routing into the vehicle fleet composition problem. Eur J Oper Res 66:313–330

Taillard ED (1999) A heuristic column generation method for the heterogeneous fleet VRP. RAIRO 33:1–34

Tarantilis CD, Kiranoudis CT, Vassiliadis VS (2004) A threshold accepting metaheuristic for the heterogeneous fixed fleet vehicle routing problem. Eur J Oper Res 152:148–158

Toth P, Vigo D (2012) The vehicle routing problem, SIAM monograph on discrete mathematics and applications. SIAM, Philadelphia, PA

Wang X, Gloden B, Wasil E (2008) Using a genetic algorithm to solve the generalized orienteering problem. In: Golden B, Raghavan S, Wasil E (eds) The vehicle routing problem: latest advances and new challenges. Springer, Berlin, 263–274

Wang Z, Li Y, Hu X (2015) A heuristic approach and a for the heterogeneous multi-type fleet vehicle routing problem with time windows and an incompatible loading constraint. Comput Ind Eng 89:162–176

Wassan NA, Osman IH (2002) Tabu search variants for the mix fleet vehicle routing problem. J Oper Res Soc 53:768–782

Yaman H (2006) Formulations and valid inequalities for the heterogeneous vehicle routing problem. Math Program 106:365–390

Chapter 9
Multi-Criterion Examination Timetabling: A Fuzzy Grouping Genetic Algorithm Approach

9.1 Introduction

The examination timetabling problem has been a challenge to the research community since the 1960s (Burke et al. 1997). Ever-increasing level of research activities on this subject has been witnessed across the globe, as evidenced by the establishment of several international research conferences such as the Practice and Theory on Automated Timetabling (PATAT) (Burke et al. 1996) and the European Association of Operational Research Societies (EURO) working group on automated timetabling (see http://www.asap.cs.nott.ac.uk/watt). Moreover, a comprehensive collection of benchmark examination timetabling problems has been established in the last two decades (Carter et al. 1996). Therefore, the problem has attracted quite a significant attention in the research community.

Educational timetabling is one of the most practical and most widely researched problems, among other timetabling problems. It is the most time-consuming and yet most important problem that periodically arises in academic institutions world-wide. Several variants of educational timetabling exist in the literature, including class-teacher scheduling (school timetabling), faculty timetabling, and classroom assignment, university course timetabling, and examination timetabling. This chapter focuses on examination timetabling.

The examination timetabling problem has a myriad of constraints (hard and soft), most of which may be conflicting. These constraints add to the complexity of the problem. Some of the hard constraints that have appeared in the literature are listed in as follows:

1. Examinations with common resources (e.g., students) cannot be assigned simultaneously; and
2. Resources of examinations must be sufficient; for instance, the size of examinations should not exceed the room capacity, and enough rooms should be available for all of examinations.

© Springer International Publishing Switzerland 2017
M. Mutingi and C. Mbohwa, *Grouping Genetic Algorithms*,
Studies in Computational Intelligence 666,
DOI 10.1007/978-3-319-44394-2_9

In addition to hard constraints, soft constraints must be satisfied in order to improve the quality of solutions. Some of the most common soft constraints are as follows:

1. Conflicting examinations should be spread as far as possible, for instance, with at least n consecutive time slots or days.
2. Some groups of examinations may preferably be at the same time, on the same day or at one location.
3. Some of the examinations may preferably be scheduled consecutively.
4. Large examinations should preferably be scheduled as early as possible.
5. Certain time slots may have preferred or limited number of students and/or examinations.
6. Examinations of the same length should be combined into the same room.

It is often desirable to develop decision methods that produce solutions that meet all hard constraints, while satisfying the soft constraints as much as possible. Efficient, interactive, and flexible multi-criterion computational approaches are most preferred. The main purpose of this chapter is to present a fuzzy multi-criterion grouping genetic algorithm for the timetabling problem. The proposed approach can model fuzzy parameters, including the intuitive choices and preferences of decision maker and management aspirations. Moreover, the approach uses unique grouping genetic operators that can exploit the group structure of the problem more efficiently.

The rest of the chapter is organized as follows: The next section briefly describes the examination timetabling problem. Section 9.3 outlines the related approaches to timetabling. Section 9.4 proposes a fuzzy grouping genetic algorithm. Numerical experiments are provided in Sect. 9.5. Section 9.6 presents the results and discussions. Section 9.7 summarizes the chapter.

9.2   The Examination Timetabling Problem

The examination timetable problem (ETP) is concerned with scheduling a set of examinations $E = \{e_1, e_2, \ldots e_m\}$ into a limited set of time slots or periods $T = \{t_1, t_2, \ldots, t_n\}$, and rooms of capacity $C = \{c_1, c_2, \ldots, c_t\}$ in each time slot, subject to hard and soft constraints. Hard constraints are supposed to be satisfied at all times, if a solution is to be feasible. For instance, a student cannot sit for two or more examinations at the same time, and the sum of students taking an examination in the same period may be limited by a prespecified capacity limit.

On the other hand, soft constraints are an expression of the wishes of the management, students, or professors. For instance, it is desirable to have large classes scheduled early to allow for early assessment of the examinations and to provide study time for each student between any two examinations, avoiding near-clashes. Near-clashes are defined by an order n ($n = 1, 2, \ldots$), where n is the number of periods between two examinations. Soft constraints may be violated, but at a cost, thus, it is highly desirable to satisfy them as much as possible.

In view of the above, a high-quality timetable is one that (i) does not violate any hard constraints and (ii) satisfies soft constraints to the highest degree possible. The ETP can be modeled from a multiple criteria perspective, based on the hard and soft constraints of the problem. There are two variants of the ETP, that is, the capacitated ETP which considers room allocation and the uncapacitated ETP which does not (Pillay and Banzhaf 2010). The uncapacitated ETP is considered in this chapter.

9.3 Related Approaches

Several solution approaches have been suggested in the literature. These approaches can be classified into low-level constructive heuristics (Burke et al. 1997; Carter et al. 1996; Pilay and Banzhaf 2010; Asmuni et al. 2005), constraint-based techniques (Freuder and Wallace 2005; Brailsford et al. 1999; Cambazard et al. 2012), decomposition techniques (Carter and Johnson 2001; Reeves 2005; De Causmaecker et al. 2009), metaheuristics (Abdullah et al. 2004, 2005, 2007, 2009, 2010, 2012; Nothegger et al. 2012; Pillay and Banzhaf 2010; Yang and Jat 2011), multi-criterion techniques (Burke et al. 2001; Cote et al. 2004), and hyper-heuristics (Kendall and Hussin 2005a, b; Burke et al. 2007; Ross 2005).

Due to the high level of complexity encountered in the timetabling problem, researchers have applied constructive heuristics which have been used since the inception of the timetabling research. Though these classical heuristics are inherently simple, their advantage is that they are easy to implement and can provide good practical solutions within reasonable computational times. In modern research applications, the techniques are widely used to incorporate into hybrid methods. Oftentimes, they are used to construct initial solutions for population-based metaheuristics and other improvement-based techniques. Among other techniques, Table 9.1 provides six well-known constructive heuristics and their brief descriptions (Burke et al. 2004; Pillay and Banzhaf 2010).

Evolutionary metaheuristics have been instrumental in modeling and solving a number of timetabling problems. Burke et al. (1994) applied a memetic algorithm to induce examination timetables, where each timetable is represented as a number of memes, one for each time slot. The algorithm combined the use of hill-climbing, and light and heavy mutation. Similarly, Ozcan and Ersoy (2005) used a memetic algorithm for a final examination scheduler system, combining genetic algorithm and violated directed hill-climbing.

Eley (2006) presented a max–min ant system (MMAS) for the uncapacitated examination timetabling problem. The hill-climbing heuristic was applied to the best timetable to further reduce the cost of each feasible timetable constructed during each cycle. It can be seen from the literature studies that a majority of the studies have used evolutionary algorithms to solve specific timetable problems from various institutions with specific constraints.

Burke et al. (1994) applied a genetic algorithm to test problems particularly for the capacitated version of the timetabling problem. In the same vein, Ross et al. (1998) applied a genetic algorithm to the capacitated version of the Carter

Table 9.1 Low-level constructive heuristics and their brief descriptions

No.	Heuristic	Strategy	References
1	Random	Randomly order the examinations	Burke et al. (1997)
2	Largest degree	Decreasing order by the number of conflicts the examinations have with the other examinations	Broder (1964)
3	Saturation degree	Order increasingly by the number of available or feasible time slots for the exam	Brelaz (1979)
4	Largest enrollment	Order decreasingly by the number of enrolments for the exam	Zeleny (1974)
5	Largest weighted degree	Order decreasingly by the weighted number of conflicts the examinations have with the other examinations	Carter et al. (1996)
6	Highest cost	Order by the cost of scheduling an examination e measured as a function of its distance from other examinations with the common students	Pilay and Banzhaf (2010)

benchmark problems. Based on the work by Falkenauer (1998), Erben (2000) used a steady-state grouping genetic algorithm for evolving examination timetables.

Multi-objective approaches to the examination timetabling problem exist in the literature. Paquete and Fonseca (2001) and Cote et al. (2004) applied multi-objective evolutionary algorithms to the examination timetabling problem, where each constraint is converted to a corresponding objective that should be satisfied by evolving the timetable. In so doing, Cote et al. (2004) presented a hybrid multi-objective algorithm (hMOEA) that maintains both a population and an archive of Pareto-optimal timetables for each generation. Efficient flexible multi-criterion algorithms are most preferred in the literature. The proposed fuzzy multi-criterion grouping genetic algorithm (FGGA) hybridizes grouping genetic algorithm, constructive heuristics, and fuzzy evaluation methods.

9.4 Fuzzy Grouping Genetic Algorithm for Multi-Criterion Timetabling

The proposed fuzzy GGA (FGGA) is a grouping genetic algorithm that utilizes one or more fuzzy controlled genetic parameters to improve the search and optimization process of the algorithm.

Figure 9.1 shows the flowchart for the proposed FGGA procedure. It starts by accepting input from the user. An initial population is then generated through constructive heuristics and then evaluated for fitness. Based on the current performance measures, including maximum fitness, average fitness, and genetic control parameter values, new adjusted parameters are derived and returned for the next generation. The iterative search and optimization continues in a loop fashion until a termination condition is satisfied. The FGGA pseudo-code is presented in Algorithm 1.

Fig. 9.1 Fuzzy grouping genetic algorithm for timetabling

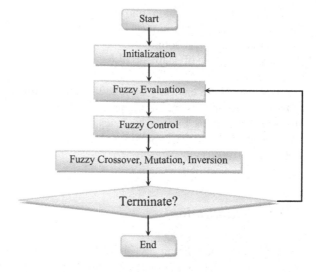

Algorithm 1 Fuzzy grouping genetic algorithm

1. **Input**: initial p^c, p^m, and p^i, population size p;
2. Initialize population $P(0)$
3. **Repeat**
4.   Fuzzy evaluation (P);
5.   Fuzzy control (P);   // classification according to the fitness
6.   Obtain fuzzy p_n^c p_n^m and p_n^i;
7.   Fuzzy crossover (), mutation (), inversion ();
8. **Until** (termination conditions is met)
9. **Return**: P

Specific elements of the FGGA procedure are presented next, including the encoding scheme, initialization, fitness evaluation, fuzzy crossover, fuzzy mutation, and fuzzy inversion.

9.4.1 Group Encoding Scheme

In implementing the FGGA, each timetable is coded using a linear structure partitioned into cell structures. Each cell represents a period in which a string of integer values are stored. The integer values encode the examinations that are allocated to that particular period. Therefore, the items are the examinations ($i = 1, 2, ..., m$), to be clustered into g groups ($j = 1, 2, ..., g$), where each group is defined by a period. Every examination i can be allocated to any group, subject to hard and soft constraints. For illustration purposes, consider a timetabling problem with 13 examinations to be partitioned into 5 groups. Figure 9.2 shows a chromosome encoding for a typical solution to this example, with groups $j = 1, 2, …, 5$, which correspond to sets of learners $\{3,6,8\}$, $\{2,5,7\}$, $\{4,9\}$, $\{10,11,13\}$, and $\{1,12\}$, respectively.

Here, each item can only be found in one group. The encoding scheme allows genetic operators to work on groups, rather than items of the chromosome. As such, the algorithm has a built-in tendency to preserve the group spirit of the coding scheme based on the basic building blocks of the chromosome (Kashan et al. 2015; Falkenauer 1996).

9.4.2 Initialization

A good initial population can significantly improve the search and optimization efficiency. As such, it is important to develop a constructive heuristic for generating the initial population, using domain-specific knowledge. The algorithm accepts input, including the available set of examination $E = \{e_1,…, e_m\}$ and periods $N = \{n_1,…, n_g\}$ into which the examinations are to be slotted. In addition, genetic parameters are also accepted as input, including the initial probabilities of crossover p_0^c, mutation p_0^m, and inversion p_0^i. After accepting the input, the algorithm assigns the most difficult examination e' to any available least cost period. Here, the cost of scheduling an examination e is calculated in terms of its distance from examinations that have the same (Pillay and Banzhaf 2010). In case of ties, priority must be given to e' with the largest student enrollment. The process is repeated until all the examinations in set E are assigned. The overall initialization algorithm is presented in Algorithm 2.

Algorithm 2 Pseudo-code for the initialization procedure

1. **Input**: Examinations $E = \{e_1,…,e_m\}$, Periods, $N = \{n_1,…,n_g\}$;
2. Order the examinations according to heuristic cost;
3. **Repeat**
4.   Select, without replacement, the most difficult unassigned exam e';
5.   Assign e' to the least cost period n';
6.   In case of ties, give priority to e' with largest enrollment;
7.   Order the remaining unassigned examinations;
8. **Until** (All examinations in E are scheduled)
9. **Return**

Fig. 9.2 The group encoding scheme for examination timetabling

Exams:	3,6,8	2,5,7	4,9	10,11,1	13,12
Group:	1	2	3	4	5

After the initialization process, the population $P(0)$ enters into an iterative loop consisting of evaluation, crossover, mutation, and inversion, until the termination condition is satisfied and is evaluated for fitness.

9.4.3 Fuzzy Evaluation

A high-quality examination timetable is expected to have no clashes, that is, no student should be scheduled to write more than one examination at any time slot. In addition to minimizing clashes, the level of proximity (or proximity cost) of examinations with common students should be minimized as much as possible. According to Carter et al. (1996), proximity cost is calculated using the following expression:

$$f_1 = \sum_{(e_i, e_j) \in C} \frac{w(|e_i - e_j|)N_{ij}}{S} \tag{9.1}$$

where C denotes the set of examinations with combinations (e_i, e_j) with common students; $|e_i - e_j|$ denotes the distance between the periods associated with each pair of examinations $(e_i, e_j) \in C$; N_{ij} is the number of students common to examinations (e_i, e_j); S is the total number of students; $w(1) = 16$, $w(2) = 8$, $w(3) = 4$, $w(4) = 2$, and $w(5) = 1$ (the smaller the distance, the higher the allocated weight).

In addition to the proximity cost, the number of clashes in a candidate timetable (or candidate solution) is used as a cost function f_2, which should ideally be reduced to zero, in practice. Let the number of clashes be defined by the following:

$$f_2 = n(e_i, e_j) \tag{9.2}$$

where $n(e_i, e_j)$ denotes the number of clashes in the examination combinations (e_i, e_j) with common students.

For the ease of computational experiments, the two functions f_1 and f_2 are normalized using a linear membership functions. To achieve this, assume that the two cost functions have an ideal (desired) minimum a, which is zero in this case, and the maximum acceptable cost b for each specific function. The values a and b, normally determined by expert opinion, are used as the parameter values for the linear membership function as shown in Fig. 9.3.

The linear membership function for this situation is defined by the following expression:

$$\eta(f) = \begin{cases} \frac{b-f}{b} & \text{If } 0 \leq f \leq b \\ 0 & \text{If otherwise} \end{cases} \tag{9.3}$$

Fig. 9.3 Linear membership
function for timetabling cost
functions

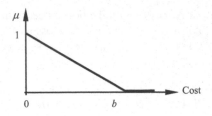

where f denotes the cost function f_1 or f_2. Therefore, the corresponding membership functions for f_1 and f_2 are as follows:

$$\mu_1 = \eta(f_1) \tag{9.4}$$

$$\mu_2 = \eta(f_2) \tag{9.5}$$

The resulting normalized function μ is obtained using the fuzzy multifactor evaluation method as follows:

$$\mu = w_1\mu_1 + w_2\mu_2 \tag{9.6}$$

where w_1 and w_2 are the respective weights of the two functions f_1 and f_2, such that $w_1 + w_2 = 1$. Since μ_1 relates to hard constraints, its weight should be much higher than that of μ_2 (for instance, $w_1 = 0.7$ and $w_2 = 0.1$). Therefore, candidate solutions are evaluated for fitness using the function μ.

9.4.4   Fuzzy Controlled Genetic Operators

The efficiency of the global iterative search process of the FGGA is influenced by the guided interaction, exploration, and exploitation. In turn, exploration and exploitation are influenced by (i) individual genetic parameters, (ii) their interaction, and (ii) the time, which affects the dynamic changes of the parameters. The magnitude of the parameter values over time plays a very crucial role on the optimization search process. These factors influence the ultimate quality of the computational results.

9.4.4.1   Fuzzy Controlled Parameters

Genetic parameters include convergence measure, diversity measure, crossover probability, mutation probability, and inversion probability. The control of their values is highly influential to the performance of FGGA. Their values are an essential input to fuzzy logic control of the search and optimization processes. Fuzzy logic control uses two major inputs:

1. The current convergence measure $M_t \in [0,1]$ or the current divergence measure $D_t \in [0,1]$, mapped into linguistic values {Low, High}; and
2. The current value of a genetic operator parameter p_0, such as crossover probability, mapped into linguistic value {Low, Medium, High}.

The output is the new genetic operator p_n, associated with linguistic values p_n = {Low, Medium, High}, to be used by the genetic operator in the next generation. Figure 9.4 shows the fuzzy control relationships and the associated membership functions. The intervals for the input, $[a_1, a_2, a_3]$ and $[b_1, b_2, b_3]$, and for the output, $[c_1, c_2, c_3]$, are obtained from expert opinion. From this analysis, specific fuzzy rules of the form **If** {Antecedent} **Then** Consequent are created.

Further details of the fuzzy controlled genetic parameters are presented next.

Convergence Measure

Convergence is a measure of the tendency of the population of chromosomes to settle down to a common solution during the optimization search process. This implies that, to obtain a good solution, premature convergence (and delayed convergence) should be avoided. In this vein, a measure of convergence M should be defined as follows:

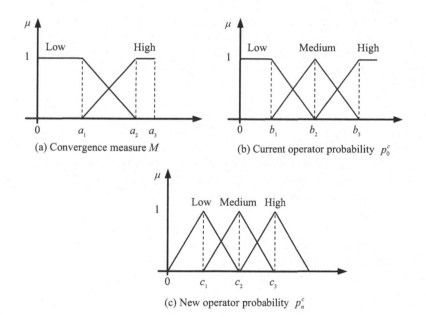

(a) Convergence measure M

(b) Current operator probability p_0^c

(c) New operator probability p_n^c

Fig. 9.4 Fuzzy membership functions for input and output of operator probability

$$M_t = \frac{f_b^c}{f_b^p} \tag{9.5}$$

where f_b^c is the fitness of the current best solution; f_b^p is the fitness of the previous best solution of the last τ generations. If M_t is high, this implies that convergence is high and most likely there was no significant improvement in the last τ generations. On the contrary, if M_t is low, this implies that the algorithm found a much better solution in the last τ generations. Because M_t is often imprecise, it usually associated with fuzzy linguistic values such as low, medium, and high.

Diversity Measure

Diversity is a measure of the closeness of chromosomes in a population at a specific generation. Like convergence measure, diversity measure can be used to control exploration and exploitation during the search and optimization process of the algorithm. Diversity D_t is be defined by the following expression:

$$D_t = \frac{f_t^{max} - f_t^{ave}}{f_t^{max}} \tag{9.6}$$

where, f_t^{max} is the maximum fitness at generation t; and f_t^{ave} is the average fitness value. Because D_t is often expressed in linguistic terms, its numerical values are mapped to a fuzzy domain, such as low, medium and high, for fuzzy control of the self-adaptive genetic parameters. It follows that if D_t is low, then convergence has taken place, and if it is high, then there is a high level of divergence in the current population.

Crossover Probability

Crossover is the most important among other genetic operators such mutation and inversion. As such, its probability (p^c) should be cautiously monitored and controlled during the global optimization process. Higher crossover probabilities promote exploration, leading to a higher rate of new solutions generated. However, further increases in p^c can be disruptive resulting in very low convergence rate and minimal exploitation. For a judicious control of the search and optimization process, the current crossover probability p_0^c is supplied as input to fuzzy control of the crossover process. A new crossover probability p_n^c is generated as output for the next generation.

Mutation Probability

The mutation mechanism is influenced by probability p^m; the higher the probability, the higher the exploitation in the neighborhoods of the solutions. However, further increase in mutation probability leads to a near random search algorithm, with little or no convergence, which may result in loss of potentially good solutions. As such, the variation of mutation probability must be controlled judiciously to guide the global optimization process. During the iterative search process, the current mutation value p_0^m is supplied as input to fuzzy logic control from which the new mutation probability p_n^m is generated. Fuzzy if-then rules are formulated in terms of p_0^m and diversity measure.

Inversion Probability

Inversion probabilistically overturns some of the candidate chromosomes so as to improve the crossover operation. However, the inversion probability p^i needs to be controlled cautiously to ensure that exploitation and exploration are balanced throughout the global optimization process. However, controlling probability p^i is difficult and imprecise; therefore, it is important to utilize expert opinion. To implement fuzzy control, the current inversion probability p_0^i and the diversity measure are used as inputs for creating a new probability p_n^i to be used in the next generation. Appropriate fuzzy membership functions and fuzzy if-then rules are constructed using expert opinion.

9.4.4.2 Fuzzy Logic Controlled Crossover

The proposed GGA utilizes a rank-based roulette wheel strategy to select parents for crossover. Each chromosome is assigned a portion of an imaginary roulette wheel, according to its rank-based fitness value. The fitness of each chromosome is evaluated according to the cost function in (9.1), and chromosomes that are fitter have higher rank-based fitness values and thus receive a relatively larger proportion of the roulette wheel. Selected chromosomes are stored in the mating pool (called *temppop*) for crossover operation.

In fuzzy-controlled crossover, the fuzzy logic controller (FLC) is fired at intervals of τ generations in order to compute a new updated value of crossover probability p_n^c, considering its value during the last τ generations and the current convergence measure. The ensuing pseudo-code for the crossover operator is presented in Algorithm 3.

Algorithm 3 Fuzzy dynamic adaptive crossover

1. **Input**: current population P, M_t, p_0^c;

2. Determine p_n^c from FLC;

3. **Repeat**

4.   Select parents, P_1 and P_2;

5.   **If** {prob p_n^c = true} **Then**   // apply two-point crossover

6.   Select crossing sections for P_1 and P_2;

7.   Interchange crossing sections of P_1 and P_2;

8.   **End**

9.   Obtain two offspring O_1 and O_2;

10.   Eliminate doubles, avoiding the crossed items;

11.   Insert misses; begin from where doubles were eliminated;

12.   Repair, if necessary;

13. **Until** (*poolsize* is reached)

14. **Return**

The crossover operation continues until the desired population of new offspring is produced.

Inputs and Outputs

There are two inputs to the FLC model: (i) the current crossover probability, p_0^c, associated with linguistic values {Low; Medium; High}, and (ii) the convergence measure M_t associated with linguistic values {Low; High}. The output p_n^c, associated with the set of linguistic values {Low; Medium; High}, is the new crossover probability to be used in the next τ generations. In real life, the intervals for the input $[a_1, a_2, a_3]$ and $[b_1, b_2, b_3]$ and the output $[c_1, c_2, c_3]$ are derived from expert opinion.

Fuzzy Rule Base

The actual fuzzy control process is determined by the FLC, which consists of a fuzzy rule base developed from expert knowledge, as illustrated in Table 9.2. In general, if f_b improves, then decrease p^c; otherwise, if no improvement in f_b, then increase p^c. This helps to ensure that premature convergence (due to very low p^c values) and genetic drift and excessive divergence (due to very high p^c values) are avoided.

Table 9.2 Fuzzy rule base for control of crossover probability p^c

Rule	M_t	p_0^c	p_n^c
1	High	Low	Medium
2	High	Medium	High
3	High	High	Low
4	Low	Low	Low
5	Low	Medium	Low
6	Low	High	Medium

9.4.4.3 Fuzzy Logic Controlled Mutation

The mutation operator fires the FLC at intervals of τ generations, where a mutation probability is adjusted according to its value p_0^m in the last τ generations and the improvement in f_b. The overall procedure for the mutation mechanism is presented in Algorithm 4.

Algorithm 4 Fuzzy dynamic adaptive mutation
1. **Input**: current mutation probability p_0^m , population P, population size p;
2. Determine p_n^m from the FLC;
2. **For** (*count* = 1 to p) **Do**
3.   **If** {prob p_n^m = true} **Then** // *Apply split and merge mutation*
4.   Select a group j' at a probability q_j
5.   Re-assign every item i from j' to any other group j at probability p_{ij}
4.   Select a group $j*$ at probability q_j, proportional group size
5.   Re-assign every item i from $j*$ to any other group j at probability p_{ij}
6.   Select two groups, $j1$ and $j2$, with probabilities q_{j1} and q_{j2}, respectively
7.   Merge the two groups, $j1$ and $j2$ into one group
8.   Repair the chromosome, if necessary
9.   **End**
10. **End**
11. **Return**

In the actual implementation, the probabilities p_{ij} and q_j are calculated from the relative sizes of the groups in a chromosome. The mutation mechanism is probabilistically performed on all the new population chromosomes.

Inputs and Outputs

The inputs to fuzzy mutation control are (i) the current mutation probability p_0^c, with linguistic values set as {Low; Medium; High}, and (ii) the convergence measure M with linguistic values {Low; High}. The output p_n^m, associated with linguistic values {Low; Medium; High}, is used for the next τ generations. As with the crossover operator, the intervals for the input $[a_1, a_2, a_3]$ and $[b_1, b_2, b_3]$ and output $[c_1, c_2, c_3]$ are derived from expert opinion.

Fuzzy Rule Base

The fuzzy rules used to control mutation are derived from the fact that if f_b nimproves, then decrease p^m; otherwise, if there is no improvement in f_b, then increase p^m. Table 9.3 further shows the fuzzy rule base for fuzzy mutation control.

The fuzzy rule base, derived from expert opinion, should ensure that p^m values are adjusted to avoid premature convergence (due to very low p^m values) and genetic drift (due to very high p^m values).

9.4.4.4 Fuzzy Logic Controlled Inversion

As in the adaptive crossover and mutation mechanisms, a new inversion probability is computed every τ generations, based on its past value in the last τ generations and the improvement achieved on best solution f_b.

Algorithm 5 The fuzzy adaptive inversion
1. **Input**: population P, current inversion probability p_0^i, population size p;
2. Determine p_n^i from FLC;
3. **For** $i = 1$ to p **Do**
4.   **If** {prob p_n^i = true} **Then** // apply the two-point inversion operator
5.   Randomly select two inversion points;
6.   Overturn the selected groups;
7.   **End**
8. **End**
9. **Return**

The inversion operator is applied at a low probability p_n^i on across the whole population.

Table 9.3 Fuzzy rule base for control of mutation probability p^m

Rule	M_t	p_0^m	p_n^m
1	High	Low	Medium
2	High	Medium	High
3	High	High	Low
4	Low	Low	Low
5	Low	Medium	Low
6	Low	High	Medium

Inputs and Outputs

Two inputs are used for fuzzy control of inversion, that is, (i) the current inversion probability, p_0^i, associated with the set of {Low; Medium; High}, and (ii) the divergence measure M_t associated with the set of linguistic values {Low; High}. The output inversion probability, p_n^i, associated with linguistic labels {Low; Medium; High}, is then used for the next τ generations. As with the previous genetic operators, the intervals for membership functions of the input $[a_1, a_2, a_3]$ and $[b_1, b_2, b_3]$ and the output $[c_1, c_2, c_3]$ are derived from expert opinion.

9.4.4.5 Fuzzy Rule Base

The fuzzy rule base for inversion control is developed based on the premise that, whenever f_b improves, the value of p^i is reduced; otherwise, whenever there is no improvement in f_b, the value of p^i is increased to promote diversity and crossover in the next generation. Table 9.4 summarizes the fuzzy rule base for the adaptive control of probability of inversion.

The above setting ensures that premature convergence (due to very low p^i values) and excessive divergence (due to very high p^i values) are always under control throughout the entire optimization process.

9.4.5 Termination

Iteratively, the algorithm loops through crossover, mutation, and inversion, until a termination condition is satisfied: (i) when the maximum number of generations, *maxgen*, is reached, or (ii) when a certain number of iterations are performed without a significant change in the current best solution, or (3) when the two conditions 1 and 2 are satisfied simultaneously. Numerical experiments and discussions are presented in the next section.

Table 9.4 Fuzzy rule base for control of inversion probability p^i

Rule	D_t	p_0^i	p_n^i
1	Low	Low	Medium
2	Low	Medium	High
3	Low	High	Low
4	High	Low	Low
5	High	Medium	Low
6	High	High	Medium

9.5 Numerical Experiments

To show the utility of the proposed approach, the computational performance of FGGA was compared to (i) other versions of genetic algorithm and (ii) other evolutionary algorithms. Comparison was based on computational efficiency in terms of computational CPU times and proximity cost. The performance of the FGGA was compared with competitive algorithms found in the literature, such as the informed genetic algorithm (IGA) developed by Pillay and Banzhaf (2010), the hybrid multi-objective algorithm (hMOEA) proposed by Cote et al. (2004), and the max–min ant system (MMAS) presented in Eley (2006). Experiments were based on the 13 well-known Carter benchmark problems listed in Table 9.5.

Preliminary computational runs were done in order to determine the appropriate genetic parameters. Table 9.6 lists the parameter values that were finally selected for further computational experiments.

9.6 Results and Discussions

The performance of the proposed multi-criterion FGGA procedure was examined based on the 13 Carter benchmark problems in the literature (Carter 2015). For each problem, 10 runs were performed, while recording the best solution obtained. The first comparative analysis was centered on computational efficiency, rated in terms of computation (CPU) times of FGGA compared to other algorithms in the literature. In particular, FGGA was compared to the informed genetic algorithm developed by Pilay and Banzhaf (2010), since this is the only available source that has performance results based on computational times.

Table 9.5 Carter benchmark problems (Carter 2015)

Problems	Examinations	Time slot	Students	Density
Car92	543	32	18,419	0.14
Car91	682	35	16,925	0.13
Ear83	190	24	1125	0.27
Hec92	81	18	2823	0.42
Kfu93	461	20	5349	0.06
Lse91	381	18	2726	0.06
Pur93	2419	42	30,029	0.03
Rye93	482	23	11,483	0.07
Sta83	139	13	611	0.14
Tre92	261	23	4360	0.18
Uta92	622	35	21,266	0.13
Ute92	184	10	2750	0.08
Yor83	181	21	941	0.29

Table 9.6 FGGA parameters for experimental runs

FGGA parameter	Value
Population size p	40
Maximum generations $maxgen$	500
Crossover probability p^c	0.45
Mutation probability p^m	0.31
Inversion probability p^i	0.08

Table 9.7 presents the computation (execution) times for the best results obtained by FGGA in comparison to those obtained from IGA developed by Pillay and Banzhaf (2010). Figure 9.5 presents a graphical comparison of the computation times. Note that the time for problem instance Pur93, as obtained by IGA, exceeds 1000 min. The graph shows that, in terms of computational times, the FGGA outperformed IGA in all but in all problem instances except three. Therefore the algorithm is computationally efficient. A closer look at the results shows that the variation of FGGA's computational times with problem type and size is not very significant, as compared to IGA. This suggests that FGGA is stable and robust over medium to large scale problem sizes.

The performance of the FGGA was also compared to other past evolutionary methodologies that were previously applied to the Carter benchmark problems. Performance rating was measured in terms of soft constraint cost of the best solution obtained by the method. These selected evolutionary algorithms include the following;

1. The two-phased approach based on an informed genetic algorithm, introduced by Pillay and Banzhaf (2010);
2. The hybrid multi-objective evolutionary algorithm (hMOEA) developed by Cote et al. (2004); and,

Table 9.7 CPU computational times (in minutes) for the best solutions obtained

Problems	FGGA	IGA[a]
Car92	53.9	71.0
Car91	66.9	98.0
Ear83	13.8	12.5
Hec92	10.9	7.5
Kfu93	46.4	52.8
Lse91	49.8	47.7
Pur93	146.6	1860.6
Rye93	52.1	72.0
Sta83	3.6	7.8
Tre92	14.6	18.7
Uta92	33.6	60.7
Ute92	3.9	11.1
Yor83	8.3	9.2

[a]IGA (Pillay and Banzhaf 2010)

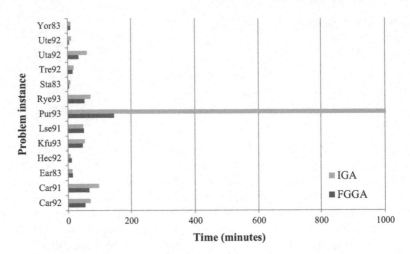

Fig. 9.5 CPU computational times (in minutes) for the best solutions obtained

3. The max–min ant system (MMAS) implemented by Eley (2006), incorporating
   the hill-climbing algorithm.

Table 9.8 lists the computational results obtained from FGGA in comparison to
other evolutionary algorithms. As can be seen from the results, FGGA outperformed other evolutionary algorithms in seven out of the thirteen problems. Thus,
the FGGA is more effective and efficient not only when compared to other versions
of genetic algorithms, but also to other related competitive evolutionary algorithms.

Table 9.8 The best results by FGGA and other evolutionary algorithms

Problems	FGGA	IGA[a]	hMOEA[b]	MMAS[c]
Car92	4.3	**4.20**	**4.20**	4.80
Car91	**4.2**	4.90	5.40	5.70
Ear83	34.6	35.9	**34.20**	36.80
Hec92	**9.8**	11.5	10.40	11.30
Kfu93	14.6	14.4	**14.30**	15.00
Lse91	**10.2**	10.9	11.30	12.10
Pur93	**4.6**	**4.70**	–	5.40
Rye93	**6.2**	9.30	**8.80**	10.20
Sta83	**136.8**	157.8	157.0	157.20
Tre92	9.6	**8.40**	8.60	8.80
Uta92	3.8	**3.40**	3.50	3.80
Ute92	28.8	27.2	**25.30**	27.70
Yor83	**32.8**	39.3	36.40	39.60

[a]IGA (Pillay and Banzhaf 2010)
[b]hMOEA (Cote et al. 2004)
[c]MMAS (Eley 2006)
Bold values indicate the best soft constraint cost obtained

However, it will be interesting to compare the algorithm with other optimization algorithms.

The performance of the FGGA approach was also compared with some of the algorithms that have made significant contributions in the field, including the following;

1. The two-phased approach based on an informed genetic algorithm, introduced by Pillay and Banzhaf (2010);
2. The sequential heuristic construction techniques proposed by Caramia et al. (2000); and,
3. The greedy randomized adaptive search procedure (GRASP) developed by Casey and Thompson (2003).

Table 9.9 lists the best final results that were obtained by the FGGA and these competitive approaches. It can be seen from the results that the algorithm managed to obtain 3 best results which outperform other algorithms. Furthermore, most of the results obtained by the algorithm are closely comparable to the selected algorithm. Therefore, the FGGA is very competitive when compared to other methods.

In view of the computational results obtained in this chapter, it will be interesting to extend the application of fuzzy multi-criterion grouping genetic algorithm to other hard problems such as cell formation problems (Onwubolu and Mutingi 2001) and bin packing problem.

Table 9.9 Comparative results from FGGA and other optimization methods

Problems	FGGA	Pillay and Banzhaf (2010)	Caramia et al. (2000)	Casey and Thompson (2003)
Car91	4.3	**4.20**	6.0	4.4
Car92	**4.2**	4.90	6.6	5.4
Ear83	34.6	35.9	**29.3**	34.8
Hec92	9.8	11.5	**9.2**	10.8
Kfu93	14.6	14.4	**13.8**	14.1
Lse91	10.2	10.9	**9.6**	14.7
Pur93	4.6	**4.70**	**3.7**	–
Rye92	**6.2**	9.30	6.8	–
Sta83	136.8	157.8	158.2	**134.9**
Tre92	9.6	**8.40**	9.4	8.7
Uta92	3.8	**3.40**	3.5	–
Ute92	28.8	27.2	24.4	25.4
Yor83	**32.8**	39.3	36.2	37.5

Bold values indicate the best soft constraint cost obtained

9.7 Summary

The examination timetabling problem is a hard problem that has attracted considerable attention of researchers and practitioners worldwide. Timetabling decision process must ensure that there are no clashes in the timetable and satisfy soft constraints as much as possible. Since the problem is highly complex, decision support systems often incorporate metaheuristic methods and domain-specific heuristics so as to address the problems more efficiently and effectively. In this vein, the chapter presented a fuzzy multi-criterion approach to model the timetabling problem. All constraints are modeled as weighted normalized cost functions using the multifactor evaluation method. The group encoding scheme adopted in this chapter enables the algorithm to capture the group structure of the problem. Enhancing fuzzy logic concepts is used to control the rate of exploration and exploitation during the search and optimization process of the algorithm.

The proposed approach contributes to the body of knowledge in the operations research and management science community. First, the suggested approach can model the fuzzy parameters of the problem, such as decision maker's choices, and preferences. Second, the approach uses unique advanced grouping genetic operators to take advantage of the group structure of the problem. Third, the approach provides a more efficient algorithm, in comparison with past approaches.

References

Abdullah S, Ahmadi S, Burke EK, Dror M (2004) Investigating Ahuja–Orlin's large neighborhood search for examination timetabling. Technical report NOTTCSTR-2004-8, School of CSiT, University of Nottingham, UK

Abdullah S, Burke EK, Mccollum B (2005) An investigation of variable neighborhood search for university course timetabling. In: The second multidisciplinary international conference on scheduling: theory and applications (MISTA), pp 413–427

Abdullah S, Burke EK, McCollum B (2007) Using a randomized iterative improvement algorithm with composite neighborhood structures for the university course timetabling problem. In: Metaheuristics. Springer, Berlin, pp 153–169

Abdullah S, Turabieh H, McCollum B, Burke EK (2009) An investigation of a genetic algorithm and sequential local search approach for curriculum-based course timetabling problems. In: Proceedings of multidisciplinary international conference on scheduling: theory and applications (MISTA 2009), Dublin, Ireland, pp 727–731

Abdullah S, Shaker K, McCollum B, McMullan P (2010) Dual sequence simulated annealing with round-robin approach for university course timetabling. In: Evolutionary computation in combinatorial optimization. Springer, Berlin, pp 1–10

Abdullah S, Turabieh H, McCollum B, McMullan P (2012) A hybrid metaheuristic approach to the university course timetabling problem. J Heuristics 18(1):1–23

Asmuni H, Burke EK, Garibaldi JM (2005) Fuzzy multiple heuristic ordering for course timetabling. In: Proceeding of the fifth United Kingdom workshop on computational intelligence. Citeseer, London, pp 302–309

Brailsford SC, Potts CN, Smith BM (1999) Constraint satisfaction problems: algorithms and applications. Eur J Oper Res 119:557–581

Brelaz D (1979) New methods to color the vertices of a graph. Commun ACM 22(4):251–256

Broder S (1964) Final examination scheduling. Commun ACM 7:494–498

Burke EK, Elliman D, Weare R (1994) A genetic algorithm based university timetabling system. In: Proceedings of the 2nd East-West international conference on computers in education, no. 1, Crimea, Ukraine, 19–23 Sept 1994, pp 35–40

Burke EK, Newall JP, Weare RF (1996) A memetic algorithm for university exam timetabling. In: Burke EK, Ross P (eds) Practice and theory of automated timetabling: selected papers from the 1st international conference. Lecture notes in computer science 1153. Springer, Berlin, pp 241–250

Burke EK, Jackson K, Kingston JH, Weare R (1997) Automated university timetabling: the state of the art. Comput J 40(9):565–571

Burke EK, Bykov Y, Petrovic S (2001) A multi-criteria approach to examination timetabling. In: Burke EK, Erben W (eds) Practice and theory of automated timetabling: selected papers from the 3rd international conference. Lecture Notes in Computer Science 2079, pp 118–131

Burke EK, Kingston JH, deWerra D (2004) Applications to timetabling. In: Gross J, Yellen J (eds) The handbook of graph theory. Chapman Hall/CRC Press, Boca Raton, pp 445–474

Burke EK, McCollum B, Meisels A, Petrovic S, Qu R (2007) A graph based hyper-heuristic for exam timetabling problems. Eur J Oper Res 176:177–192

Cambazard H, Hebrard E, O'Sullivan B, Papadopoulos A (2012) Local search and constraint programming for the post enrolment-based course timetabling problem. Ann Oper Res 194 (1):111–135

Caramia M, Dell'Olmo P, Italiano G (2000) New algorithms for examination timetabling. In: Naher S, Wagner D (eds) Algorithm engineering 4th international workshop, WAE 2000. Lecture notes in computer science 1982. Springer, Berlin, pp 230–241

Carter M (2015). ftp://ftp.mie.utoronto.ca/pub/carter/testprob

Carter MW, Johnson DG (2001) Extended clique initialisation in examination timetabling. J Oper Res Soc 52:538–544

Carter MW, Laporte G, Lee SY (1996) Examination timetabling: algorithmic strategies and applications. J Oper Res Soc 47(3):373–383

Casey S, Thompson J (2003) GRASPing the examination scheduling problem. In: Burke EK, De Causmaecker P(eds) The practice and theory of automated timetabling IV: proceedings of the 4th international conference on the practice and theory of automated timetabling, 2740. Springer, Berlin, pp 232–246

Cote P, Wong T, Sabourin R (2004) Application of a hybrid multi-objective evolutionary algorithm to the uncapacitated exam proximity problem. In: Burke EK, Trick M (eds) Practice and theory of timetabling V. 5th international conference, PATAT 2004, Pittsburgh, PA, USA, 18–20 Aug 2004, 3616. Springer, Berlin, pp 294–312 (revised selected papers). ISBN 978-3-540-30705-1

De Causmaecker P, Demeester P, Vanden Berghe G (2009) A decomposed metaheuristic approach for a real-world university timetabling problem. Eur J Oper Res 195(1):307–318

Eley M (2006) Ant algorithms for the exam timetabling problem. In 6th international conference PATAT 2006. Springer, Berlin, pp 167–180. ISBN 80-210-3726-1

Erben W (2000) A grouping genetic algorithm for graph coloring and exam timetabling. In: Burke EK, Trick M (eds) Selected papers from the third international conference on the practice and theory of automated timetabling III. Lecture notes in computer science, 2079. Springer, Berlin, pp 132–158

Falkenauer E (1998) Genetic algorithms and grouping problems. Wiley, Chichester

Falkenauer E (1996) A hybrid grouping genetic algorithm for bin packing. J Heuristics 2:5–30

Freuder EC, Wallace M (2005) Constraint programming. In: Burke EK, Kendall G (eds) Introductory tutorials in optimisation, decision support and search methodology. Springer, Berlin, pp 239–272

Kashan AH, Akbari AA, Ostadi B (2015) Grouping evolution strategies: An effective approach for grouping problems. Appl Math Modell 39(9):2703–2720

Kendall G, Hussin NM (2005a) An Investigation of a tabu search based hyper-heuristic for examination timetabling. In: Kendall G, Burke E, Petrovic S (eds) Selected papers from multidisciplinary scheduling; theory and applications, pp 309–328

Kendall G, Hussin NM (2005b) A tabu search hyper-heuristic approach to the examination timetabling problem at the MARA university of technology. In: Burke EK, Trick M (eds) Practice and theory of automated timetabling: selected papers from the 5th international conference. Lecture notes in computer science 3616, pp 199–218

Nothegger C, Mayer A, Chwatal A, Raidl GR (2012) Solving the post enrolment course timetabling problem by ant colony optimization. Ann Oper Res 194(1):325–339

Ozcan E, Ersoy E (2005) Final exam scheduler—FES. In: Proceedings of the 2005 IEEE congress on evolutionary computation, vol 2, pp 1356–1363

Onwubolu GC, Mutingi M (2001) A genetic algorithm approach to cellular manufacturing systems. Comput Ind Eng 39:125–144

Paquete LF, Fonseca CM (2001) A study of examination timetabling with multi-objective evolutionary algorithms. In: Proceedings of the 4th metaheuristics international conference (MIC 2001), pp 149–154

Pillay N, Banzhaf W (2010) An informed genetic algorithm for the examination timetabling problem. Appl Soft Comput 10:457–467

Reeves CR (2005) Fitness landscapes. In: Burke EK, Kendall G (eds) Introductory tutorials in optimization, decision support and search methodology. Springer, Berlin, pp 587–610

Ross P (2005) Hyper-heuristics. In: Burke EK, Kendall G (eds) Search methodologies: introductory tutorials in optimisation and decision support techniques, Chap 17. Springer, Berlin, pp 529–556

Ross P, Hart E, Corne D (1998) Some observations about GA-based exam timetabling. In: Practice and theory of automated timetabling II: selected papers from the second international conference, PATAT'97. Lecture notes in computer science 1408–1998. Springer, Berlin

Yang S, Jat SN (2011) Genetic algorithms with guided and local search strategies for university course timetabling. IEEE Trans Syst Man Cybern Part C Appl Rev 41(1):93–106

Zeleny M (1974) A concept of compromise solutions and method of displaced ideal. Comput Oper Res 1(4):479–496

Chapter 10
Assembly Line Balancing

A Hybrid Grouping Genetic Algorithm Approach

10.1 Introduction

An assembly line comprises a sequence of stations that repeatedly execute a given set of tasks on consecutive product units. Tasks are the indivisible elements of work to be executed at specific workstations, consuming a constant amount of time at the work station. Ideally, each unit spends the same amount of time, called the cycle time, at every station, and the reciprocal of the cycle time is the production rate. The sequence of task execution is influenced by precedence constraints which emanate from technological limitations and restrictions.

Assembly lines are widely used for mass production of complex products, even by workers with limited training (Sabuncuoglu et al. 2000). System designers seek to design efficient assembly lines with increased throughput and reduced input costs. Since planning and installation of an assembly line is a costly long-term decision problem, a careful prior assessment of the plan is essential as this will directly influence the efficiency and effectiveness of the assembly line.

In assembly line balancing, individual work elements or tasks are assigned to workstations so that the unit assembly cost is minimized as much as possible (Scholl 1999; Scholl and Becker 2006). Line balancing decisions have a direct impact on the long-run cost-effectiveness of a production process. The assembly line must be balanced and optimized before installation. As such, it is of utmost importance to develop optimal or near-optimal computational solution procedures that can assist decision makers in assembly line balancing decisions, yet with minimal computational requirements.

In spite of the importance of assembly line balance in industry, it is worth noting that the decision problem is still quite complex and computationally demanding (Sternatz 2014). In fact, the general assembly line balancing problem (ALBP) is a NP-hard combinatorial problem with complex grouping characteristics (Pape 2015; Scholl and Klein 1999; Vila and Pereira 2013). Some of the complexities include the following:

© Springer International Publishing Switzerland 2017
M. Mutingi and C. Mbohwa, *Grouping Genetic Algorithms*,
Studies in Computational Intelligence 666,
DOI 10.1007/978-3-319-44394-2_10

1. The presence of grouping structure, which is highly combinatorial and constrained;
2. The challenge of order dependency of the tasks (called items) in each work-station (called group) due to precedence constraints;
3. The challenge of order dependency of items across groups due to precedence constraints; and
4. The possibilities of problem-specific side constraints, which may be difficult to handle except by heuristics.

In light of the above issues, it is essential to develop hybrid heuristic algorithms for addressing complex ALBPs. The focus of this chapter is on the development of a hybrid grouping genetic algorithm and its application to the ALBP. In this respect, the learning outcomes for the chapter are as follows:

1. Obtain an understanding of assembly line balancing and associated complex combinatorial grouping behavior;
2. To develop and adapt the grouping genetic algorithm to the ALBP; and
3. To evaluate the strengths of the hybrid grouping algorithm in solving the ALBP.

This chapter is structured in the following manner: The next section provides a brief description of the ALBP and its grouping characteristics. This is followed by an outline of past solution approaches to the problem in Sect. 10.3. Section 10.4 presents the hybrid grouping genetic algorithm for addressing the problem. Computational experiments and results are presented and discussed in Sect. 10.5. Section 10.6 is a summary of the chapter.

10.2 Assembly Line Balancing: Problem Description

The simple assembly line balancing problem (SALBP) is quite a well-studied problem in assembly line design (Scholl 1999; Pape 2015; Jackson 1956; Johnson 1988; Gonçalves and De Almeida 2002). A set of tasks, indexed $j = 1,...,n$, are defined by processing times t_j $(j = 1,...,n)$ and their relative positions in the precedence graph. The station time S_w is equivalent to the sum of the task times of all tasks that are included in station k. The maximum station time determines the

Table 10.1 Notations for the SALBP	c	Cycle times
	j	Index of the tasks
	$J(a, b]$	Set of all tasks with $a < p_j \leq b$
	w	Index of the stations
	m	Number of stations
	n	Number of tasks
	S_w	Load of station w
	t_j	Task time of j

cycle time c, and every station w with less station time has an idle time $c - S_w$. Table 10.1 lists the notations used for the SALBP.

The aim is to find a cycle time c and a number m of stations and the corresponding task assignment which minimizes the sum of idle times over all stations, that is, $m \cdot c - \sum t_j$. In practice, a solution to the SALBP is feasible if all of the following conditions are satisfied,

1. The sum total of the task times at each station does not exceed the desired cycle time c;
2. No direct or indirect predecessor of any task j is assigned to a subsequent station assigned to j; and
3. Each task must be assigned to exactly one workstation (all tasks are non-preemptive).

It is important to note that since the objective function does not linearly depend on the variables m and c, this leads to two important variants of the SALBP, defined as follows:

- SALBP-1: Given the cycle time c, minimize the number m of stations.
- SALBP-2: Given the number m of stations, minimize the cycle time c.

Because the production rate is specified as a fixed parameter, SALBP-1 exists when a new assembly line system has to be installed and the external demand can be estimated. On the contrary, SALBP-2 holds when maximizing the production rate of an existing assembly line, which is essential for improvement in the production process. Therefore, the assembly line balancing problem seeks to determine the optimal allocation of tasks to an ordered sequence of stations, in accordance with a performance measure, such as the number of loaded workstations, line efficiency, smoothness index, balanced delay, and idle time (Chong et al. 2008; Sabuncuoglu et al. 2000; Morrison et al. 2013). A combination of the performance measures is also possible, depending on the decision focus.

Figure 10.1 presents an illustration of the SALBP problem in the form of a precedence diagram. The nodes of the precedence graph, numbered from 1 to 8, represent the tasks to be assigned to 8 workstations. Above the task nodes are their task times t_j, $j = 1,...,n$. The cycle time is assumed to be 10. The figure also identifies a feasible solution denoted by the regions encircled by dashed lines. As illustrated, the solution stipulates that task groups $\{1,2\}$, $\{3\}$, $\{4,5\}$, and $\{6,7,8\}$ are allocated to workstations 1, 2, 3, and 4, respectively. However, there are several possible combinations that may be generated from the same problem setting, which indicates the need for computationally efficient grouping algorithms for typical large-scale problems.

Several other general assembly line balancing problems are extensions of the SALBP setting; for instance, some situations may have additional space restrictions, ergonomic conditions, or mixed-model production processes. Since the inception of SALBP, several solution approaches have been suggested. An outline of the past solution approaches to the ALBP is presented in the next section.

Fig. 10.1 A basic assembly
line balancing problem

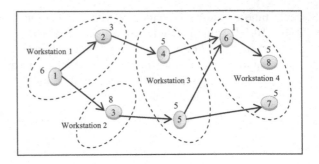

10.3   Approaches to Assembly Line Balancing

First formulated in the 1950s (Helgeson et al. 1954), the ALBP has received considerable attention from several researchers over the years (Sabuncuoglu et al. 2000; Pape 2015; Scholl and Klein 1999; Vila and Pereira 2013; Sternatz 2014; Bautista and Perreira 2002, 2007, 2009, 2011; Baybars 1986; Fleszar and Hindi 2003; Scholl and Klein 1997). Many solution methods have been suggested to solve the problem. A taxonomy of line balancing problems and their wide range of models and solution approaches are presented in Battaïa and Dolgui (2013).

Classical assembly line balancing is known as simple assembly line balancing problem (SALBP). The taxonomy of the SALBP is done according to its objective function, and some of the variants of the problem are SALBP-1, SALBP-2, SALBP-F, and SALBP-E (Scholl 1999; Scholl 1993; Scholl and Klein 1999; Chong et al. 2008; Battaïa and Dolgui 2013; Scholl and Becker 2006; Blum 2008; Boysen and Fliedner 2008; Nourie and Venta 1991).

The SALBP-1 seeks to minimize the number of workstations for a given cycle time, while SALBP-2 attempts to minimize the cycle time for a given number of workstations. On the contrary, SALBP-F checks if a feasible assembly configuration exists for a specific combination of cycle time c and number of workstations m. The SALBP-E seeks to maximize the line efficiency by simultaneously minimizing the number of workstations and cycle time.

The ALBP falls into the NP-hard class of complex combinatorial optimization problems that have been regarded as highly challenging in the production operations management research community. Consequently, several researchers shifted their attention toward heuristic methods, including the largest candidate rule (LCR) and ranked positional weight (RPW) (Scholl and Voß 1996). Though heuristics can obtain acceptable solutions at a reasonable computational cost, these approaches are too problem-specific and optimality is not guaranteed. As a result, recent research activities have shifted toward the development of efficient computational approximation algorithms or metaheuristics (Scholl and Becker 2006).

Due to the combinatorial nature of the assembly line balancing problem, metaheuristic algorithms, such as genetic algorithm (GA), tabu search (TS), simulated annealing (SA), particle swarm intelligence (PSO), and grouping genetic algorithm (GGA) are the most viable solution methods to the problem (Lapierre et al. 2006; Nearchou 2005; Ze-qiang et al. 2007). Iterative metaheuristic approaches can offer reliable solutions at reasonable computational costs. By taking advantage of the strengths of classical heuristics and metaheuristics, hybrid metaheuristics can be more promising.

10.4   A Hybrid Grouping Genetic Algorithm Approach

The hybrid grouping genetic algorithm (HGGA) incorporates the concepts of constructive heuristics and GGA to enhance intelligent iterative search of the solution space. The hybrid algorithm, though primarily developed for solving the SALBP-1 problem, can be extended to several other variants of the ALBP, with minor adjustments.

10.4.1   Encoding Scheme

The group encoding is adapted, where a feasible sequence of subsets (groups) of tasks are constructed from available tasks (items) to be performed. To enhance the iterative search and optimization process of the algorithm, it is important to ensure that the constructive heuristic is biased toward the criteria. The group structure bears an inherent meaning to the grouping problem. In this respect, the length of the chromosome is a function of the number of tasks in the precedence diagram and the number of workstations.

Figure 10.2 presents an example of group encoding scheme for a solution of a ALBP, where the terms *Items* and *Group* correspond to tasks and workstation, respectively. It can be seen that four sets of tasks, {1,2}, {3,5},{4,6}, and {7} are contained in groups 1, 2, 3, and 4, respectively. Numerous possible combinations of tasks can be generated to form such candidate solutions.

Fig. 10.2 A group encoding scheme for the line balancing problem

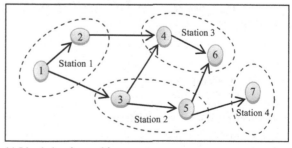

(a) Line balancing problem

Items :	1,2	3,5	4,6	7
Group :	1	2	3	4

(b) Group encoding scheme

10.4.2   Initialization

To generate a population of candidate solutions or chromosomes, a greedy constructive heuristic is developed using the station-oriented approach to assign the tasks to the workstations. The heuristic derives its concepts from the basic heuristics such as COMSOAL, RPW, and the LCR (Scholl and Voß 1996; Arcus 1966; DePuy and Whitehouse 2000; Helgeson and Bernie 1961; Otto et al. 2013, 2014; Pape 2015; Ponnambalam et al. 1999; Sprecher 1999). The RPW and LCR are used to assign tasks to workstations according to their ranked positional weights and task durations, respectively.

10.4.3   Selection

In order to obtain good offspring, performing candidates must be selected for reproduction. This is achieved through evaluation of the candidates using line efficiency as the score function;

$$f = \sum\nolimits_{j=1}^{n} (t_j) \Big/ (m - c_r) \tag{10.1}$$

where n is the total number of tasks, t_j is duration of task j, $j = 1,\ldots n$, m is the total number of workstations, and c_r is the realized cycle time. The remainder stochastic sampling without replacement is used for the selection procedure, where each chromosome k is selected for reproduction based on its expected count e_k;

$$e_k = a \cdot f_k \Big/ \sum\nolimits_{k=1}^{p} \left(\frac{f_k}{p}\right) \tag{10.2}$$

where f_k is the score function of the kth chromosome and $a \in [0,1]$ is an adjustment parameter, and each chromosome receives $[e_k]$ and additional copies through with success probability $e_k - [e_k]$. Following the section process, the candidates are set for crossover.

10.4.4   Crossover

Crossover produces a population of new offspring (selection pool), which enables the algorithm to explore unvisited regions of the solution space. Using pairs of different parent chromosomes, a two-point group crossover is repeatedly applied at probability p^c, until the selection pool size (*poolsize*) is reached. The crossover algorithm is listed below:

Algorithm 1. Crossover Operator

1. **Repeat**
2.   Select parent chromosomes, P_1 and P_2, $P_1 \neq P_2$;
3.   Select crossing sections for P_1 and P_2;
4.   Cross P_1 and P_2, obtain offspring O_1 and O_2;
5.   Eliminate *doubles*, avoiding crossed items;
6.   Insert *misses*, beginning from where doubles were eliminated;
7.   Repair the offspring, if necessary;
8. **Until** (*poolsize* is achieved)
9. **Return**

Figure 10.3 illustrates the crossover mechanism, where two randomly selected parents, $P_1 = [\{1,2\}\{4,5\}\{3,7,6\}]$ and $P_2 = [\{1,2\}\{3,5,6\}\{4,7\}]$, are crossed to produce two offspring. However, due to crossover, some items appear more than once, while others may be missing. Such offspring are repaired by eliminating repeated items (doubles) and inserting missing ones (misses), giving priority to the least loaded groups (that is workstations).

A new population (called *newpop*) is created for the next generation (or iteration) by comparing best performing offspring and the current population (called *oldpop*). The new population should then undergo mutation.

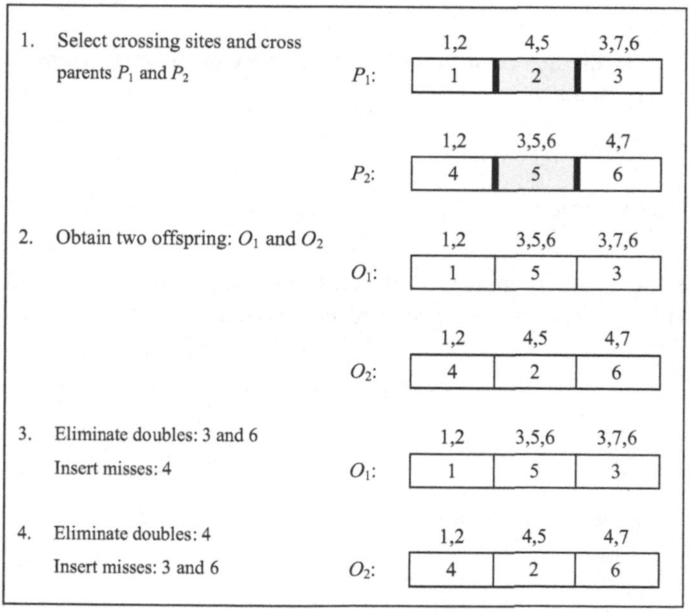

Fig. 10.3 The group crossover operation

10.4.5 Mutation

A shift mutation is adopted in this application. The mutation operator shifts items between two groups, but at a probability proportional to the relative sizes of the two selected groups. This implies that the item from the larger group is more likely to be shifted than the one from a smaller group. Mutation seeks to improve candidate solutions by probabilistic search in the neighborhood of the candidates. A repair mechanism is employed to correct candidates whose grouping violates any hard constraint.

Figure 10.4 illustrates how the shift mutation transforms chromosome [{1} {6,9,3}{2,4}{8,5,7}] to {1,5}{6,9,3}{2,4}{8,7}. The pseudo-code for the shift mutation operator is presented in the following algorithm:

Algorithm 2 Shift Mutation Operator

1. **For** *count* = 1 to p // population size = p
2.   **If** {prob (p^m) = true} **Then** mutate chromosome *count*
3.   Select two groups $j1$ and $j2$, $j1 \neq j2$;
4.   Select two items $i1$ from $j1$ and $i2$ from $j2$;
5.   Shift $i1$ to $j2$, $i2$ to $j1$, with probability p_1 and p_2, resp;
6.   Repair chromosome, if necessary;
6.   **End**
7. **End**
8. **Return**

According to the algorithm, mutation is performed subject to precedence and cycle time constraints.

10.4.6 Inversion

A two-point inversion is applied on randomly selected chromosomes, but at a very low probability p^i. The operator overturns items within two selected inversion sites.

			1	6,9,3	2,4	8,5,7
1.	Randomly select two groups:					
	1 and 4	P_1:	1	2	3	4
2.	Randomly select two items:		1	6,9,3	2,4	8,5,7
	1 and 5	P_2:	1	2	3	4
3.	Shift item 5 to smaller group 1		**1,5**	6,9,3	2,4	8,7
	(Do not shift item 1)	O_1:	1	2	3	4

Fig. 10.4 The shift mutation operator

In case of any violation of hard constraints, a repair mechanism is utilized. The algorithm for the inversion mechanism is presented as follows:

Algorithm 3 Two-Point Inversion

1. Input: p, population; inversion probability p^i;
2. **For** $i = 1$ to p **do**
3.   **If** {prob (p^i) = true} **Then**
4.   Randomly select two inversion points;
5.   Overturn groups bounded by inversion points;
6.   Repair, if need be;
7.   **End**
8. **End**
9. **Return**

The effect of the inversion operator is to improve diversity of the population of chromosomes, which in turn enhances the crossover operator. While it is desirable to maintain diversity at an acceptable diversity, it is necessary to allow the population to converge toward the end of the iterations. Therefore, the inversion probability $p^i(t)$ is subject to decay according to the following expression,

$$p^i(t) = p_0^i e^{-\alpha(t/T)} \qquad (10.3)$$

where t is the iteration or generation count; $\alpha \in [0,1]$ is an adjustment; T is the maximum number of iterations; and p_0^i is the initial probability.

10.4.7   Termination

The HGGA continues its iterative search process till a termination condition is met. Termination occurs when (1) the maximum number of iterations, T, is reached, or (2) when a certain number of iterations are performed without a significant change in the current best solution, or (3) both conditions 1 and 2 are met at the same time.

10.5   Computational Tests and Results

The proposed HGGA was coded in Java SE 8 on a Windows-based Intel Core, CPU 1.70 GHz, and 8 GB RAM computer. For each problem instance, each algorithm was run ten times, recording the best solutions for both algorithms. Computational times were also recorded for each run. The parameter settings for both algorithms were set as shown in Table 10.2.

Table 10.2 Genetic
parameters and their values

Genetic Parameter	Value
Population size, p	20
Crossover probability, p^c	0.35
Mutation probability, p^m	0.01
Inversion probability, p^i	0.04

The performance of the HGGA approach was compared to basic GA based on two criteria, that is, (i) the number of workstations and (ii) realized cycle time, in that order of priority. Thus, solutions with the lowest number of workstations are most preferable. However, in case of ties, where solutions have the same number of workstations, realized cycle time is considered. Further, the performance of the HGGA was compared to GA in terms of average computation times (or computation time).

The comparative experiments were carried out using benchmark problems from the literature (Hoffman 1963). From this data set, six problem instances were selected, considering problem size and complexity. The selection ensured a good collection of small- and large-scale problems that were obtained. The smallest instance is the Mitchell-21 with 21 tasks, while the larges is Arcus-111 with 111 tasks. Each of the problems was tested over a range of different cycle times, following the data sets in the literature (Hoffman 1963). The data sets provide a significant range of benchmark problems.

The two algorithms, HGGA and basic GA, were tested on each problem instance and for a given cycle time. In each instance, the number of workstations generated and realized cycle times were recorded, together with the average computational times.

The analysis of results was split into small-scale problems, with less than 70 tasks, and large-scale problems, with over 70 tasks. Therefore, the analysis is meant to demonstrate the performance of GA and HGGA over small- and large-scale problems.

10.5.1   Computational Results: Small-Scale Problems

Table 10.3 presents the results of the analysis of the comparative performance between hybrid GGA and the basic GA over small-scale benchmark problems, that is, Mitchell, Sawyer, and Kilbride. It is easy to see that the HGGA outperformed the basic GA in 11 out of 21 problem instances (52.39 %). For instance, consider the Mitchell-21 instance with a given cycle time of 26 time units. Though both algorithms performed equally well in obtaining 5 workstations, the HGGA performed better in terms of realized cycle time. The realized cycle time for the HGGA was 21 time units, while the one for the basic GA was 21.

Table 10.3 Comparative results between GA and hybrid GGA based on small-scale benchmark problems

Problem	Tasks	Cycle time	GA			Hybrid GGA		
			Workstations	Realized cycle time	CPU time	Workstations	Realized cycle time	CPU time
Mitchell	21	14	9	13	2.66	8	14	2.16
		15	8	15	3.98	8	15	3.62
		21	7	16	7.11	7	16	13.23
		26	5	23	13.23	5	21	11.22
		35	3	35	21.27	3	35	14.21
		39	3	39	23.17	3	36	15.14
Sawyer	30	25	14	25	10.93	14	25	9.33
		27	13	26	14.52	13	26	12.48
		30	12	30	16.85	12	29	23.69
		36	10	35	25.14	10	34	28.39
		41	8	46	31.77	8	41	30.99
		54	7	48	37.05	7	48	31.09
		75	6	58	38.98	5	66	33.95
Kilbridge	45	55	11	55	6.33	11	55	5.71
		57	10	60	9.31	10	57	6.62
		79	8	75	8.77	7	79	6.15
		92	7	81	13.64	7	81	8.14
		110	6	99	33.29	6	94	15.24
		138	5	117	38.70	4	138	25.77
		138	4	151	44.10	4	148	31.24
		184	3	184	43.16	3	184	38.94

10.5.2   Computational Results: Large-Scale Problems

As with small-scale problems, the quality of solutions is primarily determined by the number of workstations; solutions with the lowest number of workstations are preferred. In case of ties, the realized cycle time is considered.

Table 10.4 presents comparative analysis between GA and hybrid GGA based on large-scale benchmark problems (Tonge-70, Archus-83 and Arcus-111). The large-scale problems are much more complex than the small-scale problems. The first and second columns show test problems and their sizes, in terms of number of tasks. The third column indicates the cycle times for each problem. This is followed by the comparison of GA and HGGA solutions, in terms of the number of workstations, realized cycle time, as well as the CPU time.

Table 10.4 Comparative results between GA and hybrid GGA based on large-scale benchmark problems

Problem	Tasks	Cycle time	GA			Hybrid GGA		
			Workstations	Realized cycle time	CPU time	Workstations	Realized cycle time	CPU time
Tonge	70	156	25	158	23.44	25	158	19.97
		176	22	178	32.11	21	176	28.64
		195	19	206	42.30	19	194	33.79
		364	10	357	74.11	10	357	31.87
		410	9	398	68.07	9	397	68.85
		468	8	446	88.74	8	446	73.09
		527	7	506	89.09	7	506	74.82
Arcus	83	3985	20	4320	21.36	20	4098	22.01
		5048	16	4943	38.89	16	4943	27.53
		5853	14	5724	74.10	14	5621	33.69
		6842	12	6659	66.60	12	6591	58.78
		7571	11	7141	75.79	11	7141	69.51
		8412	10	8036	91.06	10	7882	87.84
		10,816	8	10,306	151.51	8	10,306	121.75
Arcus	111	5755	30	5689	56.08	27	5752	49.65
		5785	27	5796	94.12	27	5746	82.21
		8847	19	8265	90.66	18	8689	89.10
		10,027	16	9736	108.79	16	9684	112.34
		10,743	15	10,323	179.75	15	10,288	182.22
		11,378	14	11,121	193.32	14	11,121	188.76
		17,067	9	16,885	188.68	9	16,872	102.35

It can be seen that the HGGA outperformed the basic GA in 13 out of 21 problem instances (61.9 %), which is more significant than with small-scale problem instances (52.39 %). The algorithm performed better in terms of the number of workstations and/or the realized cycle time. For example, for the Tonge-70 problem with cycle of 176 time units, the basic GA obtained 22 workstations HGGA and a realized cycle time of 178, compared to the HGGA which obtained 21 workstations and a realized cycle time of 176. In addition, while the two algorithms performed equally well in terms of number of workstations on the Arcus-83 problem with a cycle time of 3985, the HGGA yielded a better realized cycle time of 4098 than 4320 obtained by the basic GA.

In view of the above analysis, it can be argued that the HGGA performance is positively influenced by the incorporation of constructive heuristics, the group encoding scheme, and the grouping genetic operators.

Table 10.5 Performance of HGGA against basic GGA over 42 experiments	Criteria	Number of instances		
		Better	Same	Worse
	No. of small-scale problems	11	10	0
	No. of large-scale problems	13	8	0
	Overall problems	24	18	0
	Percentage (%)	57.14	42.86	0

10.5.3 Overall Computational Results

Out of a total of 42 experiments, the HGGA approach outperformed the basic GA in 24 problem instances, that is, 57.14 %. However, the two algorithms performed equally well in 18 problem instances, which is about 42.86 %. A summary of the comparative performance of the HGGA approach against the basic GA is presented in Table 10.5.

In terms of average computation times, the HGGA performed much better than GA on virtually all problem instances. This is attributable to the algorithm's built-in constructive heuristics and the unique grouping genetic operators. Therefore, the HGGA is capable of solving hard large-scale assembly line balancing problems efficiently.

10.6 Summary

Assembly line balancing is a highly combinatorial and complex problem that deals with the assignment of individual work elements or tasks to workstations with the objective of minimizing the assembly cost as much as possible. Challenges associated with this problem were discussed, including the group orientation of the problem, the presence of precedence (order dependency) constraints between tasks executed within each group and across groups, and other problem-specific constraints. As a NP-hard computational problem, heuristic algorithms and meta-heuristics have been used to solve the assembly line balancing problem.

This chapter presented a hybrid grouping genetic algorithm to address complex problems. The algorithm hybridizes basic constructive heuristics, enhanced genetic operators, and other techniques to improve the optimization search process. The performance of the proposed hybrid algorithm was compared to the basic GA based on established test problems. Results of the comparative computational experiments showed that the hybrid algorithm is effective and efficient, in terms of the quality of solutions (measured by the number of workstations and the realized cycle time), as well as the average computation times.

The proposed hybrid algorithm can be developed into a decision support system to assist decision makers in making decisions associated with assembly line balancing.

References

Arcus AL (1966) COMSOAL: a computer method of sequencing operations for assembly lines. Int J Prod Res 4:259–277

Battaïa O, Dolgui A (2013) A taxonomy of line balancing problems and their solution approaches. Int J Prod Econ 142(2):259–277

Bautista J, Pereira J (2002) Ant algorithms for assembly line balancing. Ant algorithms, third international workshop, ANTS. Springer, Brussels, pp 65–75

Bautista J, Pereira J (2007) Ant algorithms for a time and space constrained assembly line balancing problem. Eur J Oper Res 177(3):2016–2032

Bautista J, Pereira J (2009) A dynamic programming based heuristic for the assembly line balancing problem. Eur J Oper Res 194(3):787–794

Bautista J, Pereira J (2011) Procedures for the time and space constrained assembly line balancing problem. Eur J Oper Res 212(3):473–481

Baybars I (1986) A survey of exact algorithms for the simple assembly line balancing problem. Manag Sci 32(8):909–932

Blum C (2008) Beam-ACO for simple assembly line balancing. INFORMS J Comput 20(4): 618–627

Boysen N, Fliedner M (2008) A versatile algorithm for assembly line balancing. Eur J Oper Res 184(1):39–56

Chong KE, Omar MK, Bakar NA (2008) Solving assembly line balancing problem using genetic algorithm with heuristics-treated initial population. In: Proceedings of the world congress on engineering 2008, vol II, WCE 2008. 2–4 July 2008, London, UK

DePuy GW, Whitehouse GE (2000) Applying the COMSOAL computer heuristic to the constrained resource allocation problem. Comput Ind Eng 38(3):413–422

Fleszar K, Hindi KS (2003) An enumerative heuristic and reduction methods for the assembly line balancing problem. Eur J Oper Res 145(3):606–620

Gonçalves JF, De Almeida JR (2002) A hybrid genetic algorithm for assembly line balancing. J Heuristics 8(6):629–642

Helgeson WB, Birnie DP (1961) Assembly line balancing using ranked positional weight technique. J Ind Eng 12:394–398

Helgeson WB, Salveson ME, Smith WW (1954) How to balance an assembly line, Technical Report, Carr Press, New Caraan, Conn

Hoffmann TR (1963) Assembly line balancing with a precedence matrix. Manag Sci 9(4):551–562

Jackson JR (1956) A computing procedure for a line balancing problem. Manag Sci 2(3):261–271

Johnson RV (1988) Optimally balancing large assembly lines with "Fable". Manag Sci 34(2): 240–253

Lapierre SD, Ruiz A, Soriano P (2006) Balancing assembly lines with tabu search. Eur J Oper Res 168(3):826–837

Morrison DR, Sewell EC, Jacobson SH (2013) An application of the branch, bound, and remember algorithm to a new simple assembly line balancing dataset. Eur J Oper Res 236(2):403–409

Nearchou AC (2005) A differential evolution algorithm for simple assembly line balancing. In: 16th International federation of automatic control (IFAC). World Congress, Prague

Nourie FJ, Venta ER (1991) Finding optimal line balances with OptPack. Oper Res Lett 10: 165–171

Otto A, Otto C, Scholl A (2013) Systematic data generation and test design for solution algorithms on the example of SALBPGen for assembly line balancing. Eur J Oper Res 228(1):33–45

Otto A, Otto C, Scholl A (2014) How to design and analyze priority rules: example of simple assembly line balancing. Comput Ind Eng 69:43–52

Pape T (2015) Heuristics and lower bounds for the simple assembly line balancing problem type 1: overview, computational tests and improvements. Eur J Oper Res 240:32–42

Ponnambalam SG, Aravindan P, Mogileeswar NG (1999) A comparative evaluation of assembly line balancing heuristics. Int J Adv Manuf Technol 15(8):577–586

Sabuncuoglu I, Erel E, Tanyer M (2000a) Assembly line balancing using genetic algorithms. J Intell Manuf 11:295–310

Scholl A (1993) Data of assembly line balancing problems. Schriften zur Quantitativen Betriebswirtschaftslehre 16/93, TU Darmstadt

Scholl A (1999) Balancing and sequencing of assembly lines. Physica-Verlag, Heidelberg

Scholl A, Becker C (2006) State-of-the-art exact and heuristic solution procedures for simple assembly line balancing. Eur J Oper Res 168:666–693

Scholl A, Klein R (1997) SALOME: a bidirectional branch-and-bound procedure for assembly line balancing. INFORMS J Comput 9(4):319–334

Scholl A, Klein R (1999) Balancing assembly lines effectively—A computational comparison. Eur J Oper Res 114(1):50–58

Scholl A, Voß S (1996) Simple assembly line balancing—Heuristic approaches. J Heuristics 2(3):217–244

Sprecher A (1999) A competitive branch-and-bound algorithm for the simple assembly line balancing problem. Int J Prod Res 37(8):1787–1816

Sternatz J (2014) Enhanced multi-Hoffmann heuristic for efficiently solving real-world assembly line balancing problems in automotive industry. Eur J Oper Res 235(3):740–754

Vilà M, Pereira J (2013) An enumeration procedure for the assembly line balancing problem based on branching by non-decreasing idle time. Eur J Oper Res 229(1):106–113

Ze-qiang Z, Wen-ming C, Tang Lian-sheng, Bin Z (2007) Ant algorithm with summation rules for assembly line balancing problem. In: 14th International conference on management science and engineering, 20–22 Aug 2007, Harbin, China, pp 369–374

Chapter 11
Modeling Modular Design for Sustainable Manufacturing: A Fuzzy Grouping Genetic Algorithm Approach

11.1 Introduction

Environmental protection has become central to every product design and processes (Ammenberg and Sundin 2005; Boothroyd et al. 1994). The use of passive approaches such as garbage classification and resource recycling can no longer cope with the increasing environmental damage. As such, it is important to develop proactive approaches to product design. Tseng et al. (2008) suggested that it is crucial to maximize the usage of resources and, at the same time, minimize the damage to the environment, as early as at the product design stage. This design approach is known as green life cycle engineering and has been discussed by a handful of authors (Otto and Wood 2001; Tseng and Chen 2004; Tseng et al. 2008). In this context, product life cycle points to the time from material usage, manufacturing, assembly, consumer usage, and the final disposal or product recycle. Here, green life cycle is centered on the last two stages, which are product use and disposal or recycle (Tseng et al. 2008). It is highly important for the designer to take a holistic look at the entire life cycle of the product in order to maximize the usage of resources and, at the same time, minimize the damage to the environment. Ideally, this should be considered as early as at the product design stage. However, it may not be possible, in practice, to precisely know the relevant design information early in the design process.

A number of researchers have explored green design from various viewpoints, including design for environment, design for recycling, and design for disassembly (Gungor and Gupta 1999; Lambert 2003; Gao et al. 2008). Modular structures have been known to improve product life cycle activities. In this vein, the concept of modularity can play a critical role in the product life cycle, in terms of improved efficiency in reuse and recycling, ease of upgrade and maintenance, ease of product diagnosis, and ease of repair and disposal, among others.

© Springer International Publishing Switzerland 2017
M. Mutingi and C. Mbohwa, *Grouping Genetic Algorithms*,
Studies in Computational Intelligence 666,
DOI 10.1007/978-3-319-44394-2_11

Though it is crucial to build green design into products at the planning stage, the idea is practically difficult to implement. The challenge is that at the design or planning stage, the information to be used for design may not be known precisely. For example, factors relating to the costs, design, and green fitness may not be precisely known early enough. Therefore, developing efficient grouping techniques is imperative (Kamrani and Gonzalez 2003; Tseng et al. 2008; Victor 2004). Thus, the purpose of this chapter was to propose a modeling approach to green modular design based on fuzzy grouping genetic algorithm.

The chapter is organized as follows: The next section gives a brief background to sustainable manufacturing. This is followed by a brief explanation of modular product design in Sect. 11.3. Section 11.4 presents a fuzzy dynamic grouping genetic algorithm approach to modular design. Section 11.5 summarizes the chapter.

11.2   Sustainable Manufacturing

Sustainable manufacturing has become a very critical concern for industry and governments across the worldwide (Seliger et al. 2008). It can be viewed as the creation of products that minimize environmental damage, minimize energy and natural resources consumption, and are economically sound. Other factors such as safety for employees, communities, and consumers are also essential. It can be seen that by definition, sustainable manufacturing must address the integration of all the three triple bottom line pillars of sustainability, namely the environmental, social, and economic indicators.

Considering the diminishing non-renewable resources, manufacturing sustainability is a very critical need. There is a need for stricter regulations on the environment and occupational safety, as consumer preference for environmentally friendly products continues to grow. Sustainable manufacturing must address issues regarding (i) economic challenges, (ii) environmental challenges, and (iii) social challenges. The green modular design approach seeks to address these issues.

11.3   Modular Product Design

Modular design is a design approach that divides a system into modules that can be independently created for assembling a variety of different systems (Gu and Sosale 1999; Gu et al. 1997; Fujita and Yosshida 2004; Huang and Kusiak 1998; Mikkola and Gassmann 2003). There are three basic categories of modular design, namely function-based modular design, manufacturing design, and assembly-based modular design (Tseng et al. 2008). Modular design is important as manufacturers need

to cope with multiple variations of product specifications and modules in a customized environment (Kreng and Lee 2004; Zha and Sriram 2006). In green life
cycle analysis, modular design is centered at the environmental level. Newcomb
et al. (1998) used group techniques to develop modular design. In the same vein,
Gu and Sosale (1999) proposed a simulated annealing algorithm for modular
design. Also, Qian and Zhang (2003) suggested a model for environmental analysis
aimed at achieving a modular goal. It can be seen from various research activities
that a well-developed modular design system will go a long way to help with the
production and control of mass customization. In addition, modular design can also
help us to focus on environmental aspects through the use of modern grouping or
clustering techniques.

In assembly-based modular design methodology, products are generally
described by liaison graph (Tseng et al. 2008). Modules are dealt with on the basis
of network partition and analysis, subassemblies, or modules (Lee 1994; Tseng
et al. 2004). There are three important stages in modular design:

1. Determination of liaison intensity of components;
2. Grouping or clustering of components using a grouping method; and
3. Evaluation of the clustering or grouping result.

An explanation of how to determine liaison intensity is presented in the literature
(De Fazio and Whitney 1987; Tseng et al. 2008; He and Kusiak 1996; Kahoo and
Situmdrang 2003). Other perspectives concerned with the evaluation of product
modularity are found in the literature (Jose and Tollenaere 2005; Tseng and Tang
2006). Figure 11.1 presents a liaison graph for Parker Pen; the nodes represent
components, and arcs represent the liaison intensity between the components.
A higher liaison intensity value indicates a close type of combination, and a smaller
value means it shows a simple type of combination.

A modeling approach based on fuzzy grouping genetic algorithm is presented in
the next section.

Fig. 11.1 A liaison graph for
Parker Pen assembly

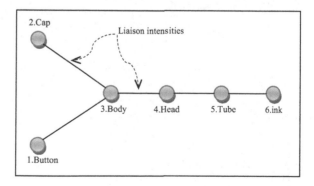

11.4 Fuzzy Grouping Genetic Algorithm Approach

Fuzzy grouping genetic algorithm (FGGA) is GGA where one of its genetic procedures incorporates fuzzy set theoretic techniques. The algorithm and its constituent components (group coding, initialization, fuzzy evaluation, and fuzzy
genetic operators) are presented.

11.4.1 Group Encoding Scheme

In this problem setting, *group* represents the modules and *items* represent the
components within a particular module. Recall the liaison graph for Parker Pen
assembly in Fig. 11.1. A typical modular product design may set six components
(items) into three modules (groups), as shown in Fig. 11.2.

The goal is to maximize the liaison intensity within each module (intra-liaison
intensity), which, in turn, minimizes the liaison intensity between modules
(inter-liaison intensity). Several combinations of these components are possible; as
the number of components and the desired number of modules increase, the number
of possible combinations increases exponentially, and the problem complexity
increases rapidly.

11.4.2 Initialization

A good initial population must be created with the hope of obtaining near-optimal
solutions within a reasonable number of iterations. It is also essential to determine
an ideal size of chromosomes, which are directly influenced by the ideal number of
modules. Ericsson and Erixon (1999) suggested a formula for the ideal number of
modules and components:

$$m = \sqrt{n} \tag{11.1}$$

where m is the number of modules and n is the number of components. Therefore,
the value of m can be used as the upper bound on the number of modules (Tseng
et al.), while the lower bound can be set at 2. Then, the initial population can be
generated using a greedy heuristic as follows:

Fig. 11.2 A group encoding
scheme for a typical modular
design

Items :	1,3,4	5,6	2
Group :	1	2	3

Algorithm 1 Initialization

1. **Input**: population size p; m_l lower bound on m; m_u upper bound on m;
2. **Repeat**
3.   Generate limits on m, Rand (m_l, m_u);
4.   Randomly assign the first component to every module;
5.   **Repeat**
6.   Assign remaining components prioritizing on liaison intensity;
7.   **Until** (all components are assigned);
8. **Until** (population size p is reached);
9. **End**
10. **Return**: Population P;

Following the initialization process, the population goes into an iterative loop consisting of evaluation, crossover, mutation, and inversion, till the termination criterion is satisfied.

11.4.3   Fitness Evaluation

In a fuzzy environment, the fitness of a chromosome must be estimated using fuzzy evaluation methods. The overall fitness function is constructed from three decision criteria: (i) the design fitness, a measure of the liaison intensity in each module, (ii) the cost analysis, consisting of material cost, manufacturing cost, and assembly cost, and (iii) green analysis, a measure of the potential environmental pollution when the product is in use. The primary objective is to maximize the liaison intensity within each module m_i, that is, intra-liaison intensity, which leads to minimization of liaison intensity across modules (inter-liaison intensity). We use the interval-valued membership functions in Fig. 11.3 to represent the fitness f, with its minimum and maximum values set at m and m' according to the expert opinion. The actual form of the fuzzy membership function depends on whether the function is to be minimized or maximized.

The three fitness functions are formulated based on the fuzzy membership functions presented.

Fig. 11.3 Fuzzy membership function for fitness function f

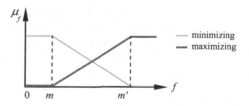

11.4.3.1   Design Fitness

The fitness of a module m_i is defined in terms of the mean accumulated liaison intensity T_i in that module (Tseng et al. 2008). Let I_{lk} be the liaison intensity between pairs of components l and k in that module. Then, the value of T_i can be estimated as follows:

$$f_i^1 = \sum_{l \in m_i} \sum_{k \in m_i} I_{lk}/2 \tag{11.1}$$

where l and k ($l \neq k$) denote the pairs of components in module m_i. The value of T_i must be maximized. For maximization, the linear membership function can be formulated as follows:

$$\mu_{f_i} = \begin{cases} 1 & f \geq m' \\ (f - m')/(m' - m) & m \leq f \leq m' \\ 0 & f \geq m \end{cases} \tag{11.2}$$

The min "∧" function is used to obtain the overall membership function μ_1 over all the modules, as follows:

$$\mu_1 = \mu_{f_1} \wedge \mu_{f_2} \wedge \cdots \wedge \mu_{f_n} \tag{11.3}$$

where n is the number of objective functions ($n = 3$ in this case). Specific values for m and m' are derived from the expert user's opinion and experiences, with emphasis on design fitness.

11.4.3.2   Cost Fitness

Cost estimation in product design can be determined from three cost considerations, namely (i) material cost, (ii) manufacturing (production) cost, and (iii) assembly cost. Therefore, the total cost can be estimated as follows (Tseng et al. 2008; Das et al. 2000; Shehab and Abdalla 2001; Zhang and Gershenson 2003):

$$f^2 = c_m + c_p + c_a \tag{11.4}$$

where c_m, c_p, and c_a are the costs associated with the material to be used, the manufacturing process, and the assembly process, respectively. For cost minimization, the membership function is expressed as follows:

$$\mu_2 = \begin{cases} 1 & f \leq m' \\ (m' - f)/(m' - m) & m \leq f \leq m' \\ 0 & f \geq m' \end{cases} \tag{11.5}$$

Specific values of m and m' are predefined by the expert user, specifically for the cost function.

11.4.3.3 Green Fitness

Green fitness can be evaluated with respect to the pollution value estimated by Eco-Indicator99 (http://www.pre.nl/), a pollution reference value which measures the potential environmental pollution caused by the component material; the higher the value, the higher the potential environmental damage when the product is in use. Therefore, the estimated pollution value is represented as follows:

$$f^3 = \sum_{k \in m_i} \omega_k p_k \tag{11.6}$$

where p_k is the unit pollution index of a component k and $_k$ is the weight (kg) of the component. The function must be minimized; therefore, the membership function can be expressed as follows:

$$\mu_3 = \begin{cases} 1 & f \leq m' \\ (m' - f)/(m' - m) & m \leq f \leq m' \\ 0 & f \geq m' \end{cases} \tag{11.7}$$

The specific values of m and m' are determined by the expert user particularly for greed design fitness.

11.4.3.4 Overall Fitness Function: Multifactor Evaluation

The resulting fitness function F_t at iteration t should also be a normalized function of its n constituent normalized functions. Therefore, F_t can be evaluated using a fuzzy multifactor evaluation method as follows:

$$F_t = \sum_{i=1}^{3} w_i \mu_i(s) \tag{11.8}$$

where $\mu_i(s)$ $(i = 1,...,3)$ denotes the constituent membership functions; w_i is the weight of μ_i; and s is a candidate solution at iteration t.

11.4.4 Fuzzy Dynamic Adaptive Operators

Fuzzy theoretic concepts are incorporated into genetic operators in order to dynamically adapt them to the state of current solutions at iteration t. The self-adaptive behavior is derived from the fact that the probabilities of crossover p^c, mutation p^m, and inversion p^i must be adjusted on real time according to the fitness of chromosomes. Also, the probabilities must be adjusted over time. For a population at iteration t, let the input variables be (i) f_{ave}, the average population fitness, (ii) f_{max}, the maximum fitness, and (iii) f_{min}, the minimum fitness of the current population. These parameters are used to control the behavior of genetic operators on real time. The parameter f_{ave} is used to divide the population into two subpopulations: (i) the weak, with fitness $f < f_{ave}$, and (ii) the fittest, with fitness $f \geq f_{ave}$. Each chromosome in each subgroup is subjected to suitable genetic operators whose parameters are adjusted according to the relative quality Q:

$$Q = \frac{f_{max} - f}{f_{max} - f_{ave}} \tag{11.9}$$

where f is the fitness of any chromosome at time t. The expression defines the relative quality of a chromosome in relation to the population. Intuitively, adaptation of the genetic operators assumes that the fittest and their offspring should survive in succeeding generations and the weak should be mutated.

11.4.4.1 Fuzzy Dynamic Adaptive Crossover

Specific crossover operators are applied to chromosomes in the two subpopulations, depending on the characteristic features of the subpopulation, which is determined by the average fitness f_{ave}. The resulting dynamic adaptive procedure is presented in Algorithm 2.

Algorithm 2 Fuzzy dynamic adaptive crossover

1. **Input:** p^c, population P
2. **Repeat**
3.   Select parents, P_1 and P_2, with respective probabilities p^{c1} and p^{c2}
4.   **If** $\{f < f_{ave}\}$ **Then**   // apply two-point crossover
5.   Select crossing sections for P_1 and P_2
6.   Interchange crossing sections of P_1 and P_2
7.   **End**
8.   **If** $\{f \geq f_{ave}\}$ **Then**   // apply single-point crossover
9.   Select two groups from P_1 and P_2
10.   Interchange the two selected groups
11.   **End**
12.   Obtain two offspring O_1 and O_2
13.   Eliminate doubles, avoiding the crossed items
14.   Insert misses; begin from where doubles were eliminated
15.   Repair, if necessary
16. **Until** (*poolsize* is reached)
17. **Return**

Note that at each crossover operation, two parent chromosomes P_1 and P_2 are selected for crossover at their respective dynamic probabilities p^{c1} and p^{c2}, whose values are controlled by the relative quality Q of each parent: (i) If Q is high, then increase p^c, maximizing the likelihood of obtaining fitter offspring, and (ii) if Q is low, then decrease p^c to avoid production of weak offspring from the weak parents. As such, p^c is adjusted accordingly as follows:

$$p^c = \begin{cases} Q \cdot k_1 e^{-t/T} & \text{If } f \geq f_{ave} \\ k_2 & \text{If } f < f_{ave} \end{cases} \tag{11.10}$$

where $k_1 = 1$ and $k_2 = 0.5$ are the characteristic control parameters crossover of the fittest and weak subpopulations, respectively. The term $e^{-t/T}$ allows the value of p^c to decrease dynamically.

11.4.4.2   Fuzzy Dynamic Adaptive Mutation

Specific mutation operators are applied to each subpopulation, depending on the characteristic fitness of the subpopulation. For a population of size p, the following algorithm is applied:

Algorithm 3 Fuzzy dynamic adaptive mutation

1. **Input**: population P; population size p
2. **For** (*count* = 1 to p) **Then**
3.   **If** {{$f < f_{ave}$} **Then**   // *Apply split and merge mutation*
4.   Select a group j' at a probability q_j
5.   Re-assign every item i from j' to any other group j at probability p_{ij}
4.   Select a group j^* at probability q_j, proportional group size
5.   Re-assign every item i from j^* to any other group j at probability p_{ij}
6.   Select two groups, $j1$ and $j2$, with probabilities q_{j1} and q_{j2}, respectively
7.   Merge the two groups, $j1$ and $j2$ into one group
8.   Repair the chromosome, if necessary
9.   **End**
10.   **If** {$f \geq f_{ave}$} **Then**   // apply swap mutation
11.   Randomly choose two different groups
12.   Randomly select two items, one from each group.
13.   Swap the selected items, and repair the resulting chromosome
14.   **End**
15. **End**
16. **Return**

Based on the relative quality Q, the mutation probability p^m is controlled in this manner: (i) If Q is high, then decrease p^m to avoid mutation of the fittest chromosomes, and (ii) if Q is low, then increase p^m to promote mutation of the weak chromosomes. Thus, p^m is adjusted as follows:

$$p^m = \begin{cases} Q \cdot k_3 e^{-t/T} & \text{If } f \geq f_{ave} \\ k_4 & \text{If } f < f_{ave} \end{cases} \tag{11.11}$$

where $k_3 = 0.7$ and $k_4 = 0.4$ define the characteristic control parameters for mutation of the subpopulations, respectively. The term $e^{-t/T}$ causes p^m to decrease dynamically.

11.4.4.3 Fuzzy Dynamic Adaptive Inversion

The inversion operator overturns genes of chromosomes at a low probability p^i, in order to promote diversity. Specific inversion operators are applied to specific subpopulations, according to the features of each subpopulation and the inversion operators. As a general heuristic rule,

Algorithm 4 The fuzzy adaptive inversion

1. **Input**: population P; population size p;
2. **For** $i = 1$ to p **Do**
3.   **If** $\{f < f_{ave}\}$ **Then** // apply the two-point inversion operator
3.   **If** {prob p^i = true} **Then**
4.   Randomly select two inversion points;
5.   Overturn the selected groups;
7.   **End**
8.   **End**
9.   **If** $\{f \geq f_{ave}\}$ **Then** // apply the full inversion operator
3.   **If** {prob p^i = true} **Then**
4.   Overturn all the groups;
7.   **End**
8.   **End**
6.   Repair, if need be;
10. **End**
11. **Return**

As can be realized from the algorithm, two-point inversion overturns two genes of a chromosome, while full inversion overturns the entire chromosome. The value of Q controls the value of p^i: (i) If Q is high, then increase p^i, so as to promote crossover of the fittest chromosome, and (ii) if Q is low, then decrease p^i to minimize crossover of weak chromosomes. Therefore, the p^i values are adjusted as follows:

$$p^i = \begin{cases} Q \cdot k_5 e^{-t/T} & \text{If } f \geq f_{\text{ave}} \\ k_6 & \text{If } f < f_{\text{ave}} \end{cases} \qquad (11.12)$$

where $k_5 = 0.7$ and $k_6 = 0.5$ define the characteristic control parameters for inversion of the fittest and weak subpopulations, respectively. As with the previous operators, the term $e^{-t/T}$ causes p^i to decay dynamically over time t.

11.4.5 Termination

The FGGA search and optimization continues iteratively until a termination condition is satisfied: when (1) prespecified maximum number of iterations, T, is reached, or (2) when the current best solution has not improved in a prespecified number of iterations, or (3) when the two conditions 1 and 2 are satisfied simultaneously.

11.5 Summary

Green modular design is one of the most effective techniques for promoting sustainable manufacturing. A good green modular design calls for a good choice of combinations of components and modules of a design, aimed at maximizing design fitness, while minimizing the product cost and potential environmental damage by the product when in use. It was realized that it is important to build manufacturing sustainability into the whole life cycle of a product (or process), which takes into account the ultimate potential impact of the product on the economic, environmental, and the social performance of the product/process. However, in a fuzzy environment, most of the necessary design information is not precisely known at the planning or design stage. Moreover, the combinatorial problem is computationally challenging, demanding more efficient and effective optimization methods. In this paper, a multiple criteria grouping approach was suggested for evaluating possible modular designs.

This chapter proposed a modeling approach for modular design based on fuzzy grouping genetic algorithm. The proposed modular design approach is promising, especially when the design factors and the criteria for evaluation, such as design fitness, cost fitness, green fitness, and other relevant information, are not precisely known at the design stage. Fuzzy evaluation techniques are used to express the imprecise information for evaluating the fitness of potential solutions in the iterative genetic algorithm. Furthermore, the algorithm uses fuzzy-based control techniques to dynamically adapt genetic parameters of the algorithm on real time.

References

Ammenberg J, Sundin E (2005) Product in environmental management systems: drivers, barriers and experiences. J Clean Prod 13:405–415

Boothroyd G, Dewhurst P, Knight W (1994) Product design for assembly and manufacture. Marcel Dekker, New York

Das SK, Yedlarajiah P, Narendra R (2000) An approach for estimating the end-of-life product disassembly effort and cost. Int J Prod Res 38(3):657–673

De Fazio DF, Whitney DE (1987) Simplified generation of all mechanical assembly sequences. IEEE J Robot Automat 3(6):640–658

Ericsson A, Erixon G (1999) Controlling design variants: modular product platforms. ASME Press, New York

Fujita K, Yosshida H (2004) Product variety optimization simultaneously designing module combination and module attributes. Concur Eng Res Appl 12(2):105–118

Gao F, Xiao G, Chen JJ (2008) Product interface reengineering using fuzzy clustering. Comput-Aided Des 40(4):439–446

Gu P, Hashemian M, Sosale S (1997) An integrated modular design methodology for life-cycle engineering. Ann CIRP 46:71–74

Gu P, Sosale S (1999) Product modularization for life cycle engineering. Robot Comput Integr Manuf 15:387–401

Gungo A, Gupta SM (1999) Issues in environmentally conscious manufacturing and product recovery: a survey. Comput Ind Eng 36:811–853

He DW, Kusiak A (1996) Performance analysis of modular products. Int J Prod Res 34(1): 253–272

Huang CC, Kusiak A (1998) Modularity in design of products and systems. IEEE Trans Syst Man Cybern—Part A: Syst Hum 28(1): 66–77

Jose A, Tollenaere M (2005) Modular and platform for product family design: literature analysis. J Intell Manuf 16:371–390

Kamrani AK, Gonzalez R (2003) A genetic algorithm-based solution methodology for modular design. J Intell Manuf 4(6):599–616

Kahoo LP, Situmdrang TD (2003) Solving the assembly configuration problem for modular products using immune algorithm approach. Int J Prod Res 41(15):3419–3434

Kreng VB, Lee TP (2004) Modular product design with grouping genetic algorithm—a case study. Comput Ind Eng 46:443–460

Lambert AJD (2003) Disassembly sequencing: a survey. Int J Prod Res 41:3721–3759

Lee K (1994) Subassembly identification and evaluation for assembly planning. IEEE Trans Syst Man Cybern 24(3):493–502

Mikkola JH, Gassmann O (2003) Managing modularity of product architecture: toward an integrated theory. IEEE Trans Eng Manage 50:204–218

Newcomb PJ, Bras B, Rosen DW (1998) Implications of modularity on product design for the life cycle. Trans ASME 120:483–490

Otto K, Wood K (2001) Product design—technical in reverse engineering and new product development. Prentice Hall, London

Qian X, Zhang HC (2003) Design for environment: an environmental analysis model for the modular design of products. In IEEE international symposium on electronics and the environment, Boston, pp 114–119)

Seliger G, Kim H-J, Kernbaum S, Zettl M (2008) Approaches to sustainable manufacturing. Int J Sustain Manuf 1(1/2):58–77

Shehab E, Abdalla H (2001) Manufacturing cost modeling for current product development. Rob Comput-Integr Manufact 17:341–353

Tseng H-E, Chang C-C, Li J-D (2008) Modular design to support green life-cycle engineering. Expert Syst Appl 34:2524–2537

Tseng HW, Chang TS, Yang YC (2004) A connector-based approach to modular formulation problem for mechanical products. Int J Adv Manufact Technol 24:161–171

Tseng HW, Chen WS (2004) A replacement consideration for the end-of-life product in the green life cycle environment. Int J Adv Manufact Technol 24:925–931

Tseng HW, Tang CE (2006) A sequential consideration for assembly sequence planning and assembly line balancing using the connector concept. Int J Prod Res 44(1):97–116

Victor BK (2004) Modular product design with grouping genetic algorithm—a case study. Comput Ind Eng 46(3):443–460

Zha XF, Sriram RD (2006) Platform-based product design and development: a knowledge-intensive support approach. Knowl Based Syst 19:524–543

Zhang Y, Gershenson JK (2003) An initial study of direct relationships between life-cycle modularity and life-cycle cost. Concur Eng Res Appl 11(2):121–128

Chapter 12
Modeling Supplier Selection Using Multi-Criterion Fuzzy Grouping Genetic Algorithm

12.1 Introduction

Supplier evaluation and selection are one of the most critical activities in most industry disciplines. Its main purpose is to streamline material/service flows, reduce manufacturer and supplier costs, and improve quality and customer service (Weber 1991). This usually calls for a precise assessment of the relevant strengths and weaknesses of suppliers. Thus, the supplier selection process consists of two major steps: (i) performance evaluation of suppliers and (ii) final selection. For a small set of potential suppliers and a simple selection criterion, the process is trivial. However, in the real world, a number of complicating features occur, so much that the decision process becomes a more complex undertaking. In fact, the decision processes often involve many conflicting criteria that have to be optimized simultaneously. Some of the most common criteria are price, lead time (delivery), quality, and the number of suppliers or vendors selected. The first three have received a considerable attention in the literature, far more than other criteria (Weber 1991).

When assessing or evaluating potential suppliers and contractors, however, the presence of fuzzy and multiple criteria may complicate the whole evaluation process. In most industry settings, the decision process often involves conflicting management goals, multiple criteria, and a number of constraints. For example, when selecting subcontractors in construction industry, the execution of the tasks may be subject to complicating restrictions, such as due dates and precedence constraints. Moreover, some situations may require that the number of suppliers or contractors should be minimal, in which case some of them have to be assigned a group of tasks or deliveries. Such approaches are most favorable, in line with the concept of the economies of scale.

Recently, several other criteria have been pointed out, yet have received very little attention. Warranties, claim policies, reputation, performance history, labor relations record, and amount of past business, and other concerns such as packaging ability and training aids are worth considering. Most, if not all, of these criteria are

© Springer International Publishing Switzerland 2017   213
M. Mutingi and C. Mbohwa, *Grouping Genetic Algorithms*,
Studies in Computational Intelligence 666,
DOI 10.1007/978-3-319-44394-2_12

qualitative, subjective, or imprecise in nature. Consequently, it is very difficult to model these criteria in precise terms. Where several entities and variables are to be taken into account, such as the number of items to be supplied, the number of tasks to be subcontracted, and the number of potential suppliers to be evaluated, the selection decision becomes even more challenging. When properly modeled and solved, the resulting decisions are expected to be far reaching in satisfying the overall business objectives. In summary, complicating features often faced by decision analysts in supplier/contactor selection include the following:

1. A number of conflicting selection criteria may need to be considered in the optimization process;
2. Most of the management goals and aspiration criteria may be fuzzy and conflicting;
3. A number of potential suppliers may need to be evaluated before the final decision can be made; and
4. The presence of complicating constraints such as due dates and precedence constraints adds to the complexity of the problem.

In light of the above issues, it will be useful to develop methodologies that can handle qualitative, imprecise, or fuzzy criteria, conflicting management goals, and complicating constraints. Moreover, new methodologies are expected to handle large-scale problems, especially given that problem sizes are ever-growing as supply networks continue to grow in modern business. Such complex situations demand efficient, adaptive, and interactive algorithms that can handle fuzzy goals, constraints, and other system parameters. Existing methodologies do not seem to have abilities to address these situations. This chapter seeks to fill these gaps and voids. This chapter presents a fuzzy multi-criterion modeling approach for addressing supplier selection problems from a fuzzy multi-criterion perspective. The proposed multi-criterion algorithm utilizes fuzzy multifactor evaluation methods to model multiple criteria by converting constraints, management goals, and management aspirations into normalized fuzzy membership functions. Clarifications and other illustrative examples are provided based on typical examples.

The rest of this chapter is organized in this manner: The next section gives a brief review of the literature related to the supplier selection methodologies. Section 12.3 provides an illustrative example on subcontracting industry. A multi-criterion fuzzy grouping algorithm approach is proposed in Sect. 12.4. Finally, Sect. 12.5 summarizes this chapter.

12.2 Related Literature

A number of methods have been proposed for solving supplier or subcontractor selection problems in the literature. These methods can be classified into 5 categories, namely: (1) basic heuristic methods, (2) mathematical programming methods, (3) Metaheuristic optimization methods, and (4) multi-criterion decision-making methods, and (5) hybrid methods which combine two of more of the above methods. Among the five categories, the fifth looks more promising for complex large-scale problems than the rest. In the first category, basic rules are used to first eliminate suppliers who do not satisfy defined selection rule which defines a minimal score (Wright 1975; Yahya and Kingsman 1999). Thus, only suppliers whose marks satisfy the minimum mark of all the selected criteria are considered for the next phase. Lastly, the supplier who satisfies the selected criterion much better than the rest of the suppliers is finally selected (Rankovic et al. 2011).

The second category uses mathematical programming techniques such as integer programming, goal programming, and dynamics programming, as well as their variants. Applications of these methods are found in the literature (Choudhary and Shankar 2014; Jadidi et al. 2015; Rajan et al. 2010).

The third category makes use of global optimization techniques to model large-scale problems. These methods include evolutionary algorithms and local search methods such as genetic algorithms, particle swarm optimization, and tabu search. The approaches are designed to handle single criterion or multiple criteria supplier selection problems subject to constraints. Several applications of these methods are found in a wide range of literature (Yang et al. 2011; Cheng et al. 2011; Polat et al. 2015).

The fourth category uses multiple criteria evaluation, including fuzzy set theoretic techniques, to model multiple conflicting criteria common in supplier selection problems. Common examples of multiple conflicting criteria are cost, quality, safety, and conveniences, which are often fuzzy in nature. Fuzzy multi-criterion evaluations are essential for decision making.

Hybrid methods are a combination of two or more of the known approaches. The aim of hybridizing these methods is to take advantage of the strengths of the methods while eliminating their weaknesses. A number of these methods have been applied in solving complex problems (Chen et al. 2006). Table 12.1 gives a taxonomy of the approaches found in the literature.

Hybrid methods are a combination of two or more of the known approaches. The aim of hybridizing these methods is to take advantage of the strengths of the methods while eliminating their weaknesses. A number of these methods have been applied in solving complex problems (Chen et al. 2006).

In the case of multiple criteria optimization and multiple constraints, the problem becomes complex (Polat et al. 2015). Furthermore, when management goals and aspirations are uncertain and imprecise, the problem becomes even more complex to handle using conventional methods. Consequently, the use of fuzzy multi-criterion optimization approaches becomes the most viable and practical methodology (Mutingi and Mbohwa 2016; Vijay and Ravindran 2007; Weber et al. 1991). For illustration purposes, we present an example of fuzzy multi-criterion optimization in subcontractor selection, a problem which is commonplace in construction industry.

Table 12.1 A taxonomy of approaches to supplier selection methods with examples

Category	Selected examples	Reference
1. Heuristic methods	Vendor rating using analytic hierarchy process method	Yahya and Kingsman (1999)
	Supplier selection in automobile industry using balanced scorecard and fuzzy analytical hierarchical process approach	Galankashi et al. (2016)
2. Mathematical programming methods	Supplier selection, carrier selection, and inventory lot-size using a goal programming model	Choudhary and Shankar (2014)
	Supplier selection using a multi-choice goal programming approach for problems	Jadidi et al. (2015)
	Integer linear programming model for supplier selection in a two-stage supply chain	Rajan et al. (2010)
3. Metaheuristic optimization methods	Genetic algorithm for solving a multi-product supplier selection model with service level and budget constraints	Yang et al. (2011)
	Subcontractor selection using evolutionary fuzzy hybrid neural network	Cheng et al. (2011)
	Subcontractor selection using genetic algorithm	Polat et al., 2015
4. Multi-criterion decision approaches	Fuzzy clustering method based on a new distance for interval type-2 fuzzy sets	Heidarzade et al. (2016)
	Green supplier selection using fuzzy group decision-making methods	Banaeian et al. (2016)
	A fuzzy multi-criterion group decision-making approach using quality function deployment, fuzzy information, and 2-tuple linguistic representation model	Karsak et al. (2015)
5. Hybrid Methods	Multi-criterion supplier selection and order using multi-criterion decision analysis and linear programming	Sodenkamp et al. (2016)
	A hybrid group decision support system using analytic hierarchy process, fuzzy set theory, and neural network	Kar (2015)
	Supplier selection using an integrated analytic hierarchy process and linear programming	Ghodsypour and O'Brien (1998)

12.3 A Subcontractor Selection Example

Selecting subcontractors out of several potential contractors is a very common problem in construction and the related industry (Polat et al. 2015; Hartmann et al. 2009). In most cases, construction management agencies are faced with the challenge of coordinating and controlling several subcontractors, where each subcontractor is responsible for a set or group of tasks. Polat et al. pointed out several reasons for subcontracting in construction projects, including (1) complexity of construction projects, (2) the need for specialized skills in certain projects, (3) the need for highly specialized equipment, (4) the need for utilization of the limited time by subcontracting specialists, (5) the need for high quality work offered by specialized contractors, and (6) to share risks with subcontractors.

When selecting multiple subcontractors out of several potential subcontractors, the aim is to select the most appropriate set of subcontractors and assign each one of them a specific group of tasks in the best possible way. In the case of small-scale problems with a few subcontractors, a single criterion, and a few constraints, the problem may be trivial. However, for large-scale problems with fuzzy and some-times conflicting multiple criteria, conventional approaches such as linear pro-gramming and basic heuristic approaches may not be able to address such problems. In the presence of multiple selection criteria, such as cost, quality, and delivery time, the problem becomes even more complex. In such situations, fuzzy grouping multi-criterion approaches methodologies are most appropriate (Mutingi and Mbohwa 2014).

Consider an example of a subcontractor selection problem where 3 subcon-tractors are available for a set of 10 tasks T = {T1, T2,...,T10}. Each task has a specific estimated execution time (duration) with respect to each subcontractor. The tasks have precedence constraints as shown in the network diagram in Fig. 12.1. Each subcontractor specifies a bid price for each task. In addition, each subcon-tractor is rated for performance quality and delivery time, as shown in Table 12.2. There are several combinations or groups of tasks or contract job that can be formed and assigned to subcontractors, subject to precedence constraints, for instance:

T1-T3-T4-T8
T1-T3-T6
T1-T2-T4
T3-T5-T10
T4-T9-T10

Each group of tasks can be assigned to a specific subcontractor, but at a specific cost. Thus, the problem is a typical grouping problem. The aim is to obtain a

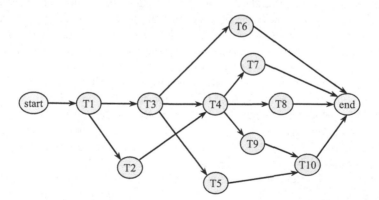

Fig. 12.1 Network diagram for task precedence

Table 12.2 Table performance of potential subcontractors in terms of delivery time, prices, and quality

Task	Contractor 1			Contractor 2			Contractor 3		
	Time (days)	Price (US$)	Quality (%)	Time (days)	Price (US$)	Quality (%)	Time (days)	Price (US$)	Quality (%)
T1	8	890,500	72	9	850,000	74	9	900,000	75
T2	5	110,500	78	6	120,500	78	7	100,500	84
T3	10	115,000	75	8	110,000	81	9	110,000	80
T4	4	65,000	85	5	80,000	88	6	60,000	72
T5	12	315,000	95	10	300,000	76	8	310,000	65
T6	8	198,000	76	9	175,000	77	8	180,000	60
T7	4	85,500	81	6	80,500	89	5	80,500	66
T8	10	310,000	70	8	290,000	91	9	320,000	80
T9	10	310,000	76	10	312,000	92	11	305,000	81
T10	2	75,000	82	3	78,000	70	4	70,000	90

combination of groups of tasks for each supplier such that the overall cost is minimized. The presence of precedence constraints implies that the sequence of tasks is essential, which makes the grouping problem order-dependent. In some cases, the number of subcontractors has to be minimized. The problem can be modeled from a fuzzy multi-criterion grouping genetic algorithm perspective.

12.4 A Fuzzy Multi-Criterion Grouping Genetic Algorithm

Fuzzy grouping genetic algorithm (FGGA), is a development from the grouping genetic algorithm (GGA) (Falkenauer 1992, 1996), wherein one or more operators of the algorithm are controlled by or use fuzzy parameters. For example, in this application, the evaluation and selection operators use a fuzzy multifactor evaluation approach. Figure 12.2 shows the flowchart for the FGGA. The algorithm starts by obtaining input such as crossover, mutation, inversion probabilities, and population size. An initial population of chromosomes (candidate solutions) is generated randomly through user-generated seeds or other constructive heuristics. The population is then evaluated for fitness. Performing chromosomes are selected from the parent population *oldpop* into a mating pool *tempop* for crossover operation by which new and possibly better performing offspring are produced, called selection pool *spop*. A new population (*newpop*) is then formed by combining best performing chromosomes from *oldpop* and *spop*. The *newpop* goes through the mutation and inversion operators by which the chromosomes are mutated and diversified at low probabilities. The resulting population is fed back to the fuzzy evaluation operator for further evaluation. The iterative cycle continues until a termination condition is satisfied, and the final population of candidate solutions is obtained.

Fig. 12.2 Fuzzy grouping genetic algorithm methodology

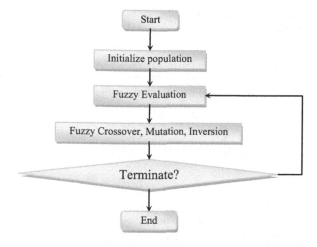

The FGGA procedure is summarized according to the pseudo-code in Algorithm 1.

Algorithm 1 Fuzzy grouping genetic algorithm
  1. **Input:** initial p^c, p^m, and p^i, population size p;
  2. Initialize population $P(0)$
  3. **Repeat**
  4.   Fuzzy evaluation (P);
  5.   Crossover ();
  6.   Mutation ();
  7.   Inversion ();
  6. **Until** (termination criteria is met)
  7. **Return:** P

The specific stages of the multi-criterion FGGA procedure are presented and illustrated further.

12.4.1 FGGA Coding Scheme

Usually, FGGA and GGA have similar coding schemes. The group coding scheme ensures ease of information coding and decoding as the iterative search process proceeds. The subcontractor selection problem in Sect. 12.3 provides a good example. Since the problem has precedence restrictions, the coding should be order-dependent. This implies that the sequence of tasks or jobs in each group should take into account the precedence constraints associated with the tasks in the group. If the group size is limited, then the length of the chromosome should be fixed. In accordance with the subcontractor selection example, the length of the chromosome should be fixed to three.

Figure 12.3 presents the group coding for our subcontractor selection problem. In this case, the items represent the tasks clustered into each group, and each group is identified by the subcontractor to which the tasks are assigned. Thus, groups of tasks {1,3,4,8}, {2,5,9,10}, and {6,7} are assigned to subcontractor 1, 2, and 3, respectively. It can be seen that the sequence or order of tasks observes the precedence constraints depicted by Fig. 12.1 in the previous section.

12.4.2 Initialization

An efficient initialization algorithm should consider hard constraints such as precedence restrictions and group size constraints. An efficient initialization procedure can significantly improve the computational efficiency of the overall algorithm. For the subcontractor selection problem, the use of a constructive heuristic or user-defined seeds is beneficial. User-generated seeds are generated by the expert user, deriving from the structure of the problem. The expert user is expected to have, through experience, appreciable prior knowledge of good candidate solutions. To avoid premature convergence, it is important to ensure that the population has an acceptable level of diversity. After initialization, population chromosomes are evaluated for fitness using fuzzy multi-criterion evaluation.

12.4.3 Fuzzy Fitness Evaluation

Real-world subcontractor problems are often characterized with conflicting fuzzy multiple criteria. As such, fuzzy multifactor evaluation functions are the best option for fitness function evaluation. Assume a typical situation with n objective functions over a solution space R, where each objective function $f_i(s)$, $(i = 1,\ldots,n)$, $s \in R$, is mapped into a corresponding normalized fuzzy function $\mu_i(s)$. It follows that the overall fitness, F_t, of a candidate solution at iteration t can be expressed as a function of its constituent normalized functions $\mu_i(i = 1,\ldots,n)$, as follows:

$$F_t = \sum_{i=1}^{n} w_i \mu_i(s) \qquad (12.1)$$

where n is the number of normalized functions $\mu_i(s)$ $(i = 1,\ldots,n)$; w_i is the corresponding weight of μ_i, such that $\sum w_i = 1.0$; and s is a candidate solution.

Fig. 12.3 The FGGA chromosome coding scheme

Learners:	1,3,4,8	2,5,9,10	6, 7
Group:	1	2	3

For our subcontractor example with 3 subcontractors and 10 tasks, the aim is to determine the most satisfactory combination of subcontractors that minimizes project duration f_1 and the total cost f_2, while maximizing quality f_3. The three functions are normalized using the fuzzy membership functions presented in Fig. 12.4.

For the case of minimization, the fuzzy membership functions μ_1 and μ_2 corresponding to f_1 and f_2 are determined by the expression,

$$\mu = \begin{cases} 1 & \text{If } 0 \geq f \leq a \\ (a-f)/(a-b) & \text{If } a \leq f \leq b \\ 0 & \text{Otherwise} \end{cases} \qquad (12.2)$$

where a and b are the parameters for the fuzzy membership function μ, which represents μ_1 and μ_2. For the maximization case, f_3 is normalized to μ_3, which is determined by the following expression:

$$\mu_3 = \begin{cases} 1 & \text{If } f_3 \geq b \\ (a-f_3)/(a-b) & \text{If } a \leq f_3 \leq b \\ 0 & \text{Otherwise} \end{cases} \qquad (12.3)$$

where a and b are the parameters for the fuzzy membership function μ_3. It is important to note that, in practice, parameters a and b correspond to the minimum and maximum conceivable values of each of the objective function, f_i, $(i = 1,\ldots,n)$. Such values can be derived from expert experience and opinion, and management goals and aspirations.

12.4.4   Selection

Well-known selection strategies, include deterministic sampling, remainder stochastic sampling with/without replacement, and stochastic tournament, have been suggested in the literature (Mengshoel and Goldberg 1999; Kashan et al. 2015; James et al. 2007). In this application, a variant of the remainder stochastic sampling without replacement is suggested, where each chromosome k is selected for

Fig. 12.4 Fuzzy membership function for function f

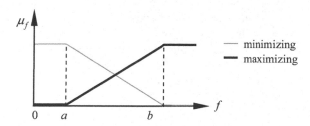

crossover in proportion to its expected count e_k determined by the following expression:

$$e_k = a \cdot F_k \bigg/ \sum\nolimits_{k=1}^{p} \left(\frac{F_k}{p} \right) \tag{12.4}$$

where F_k is the score function of the kth chromosome and $a \in [0,1]$ is a user-defined adjustment parameter. By this approach, each chromosome obtains $[e_k]$ copies and probabilistically receives additional copies at a success probability equivalent to the fractional part of e_k.

12.4.5 Adaptive Crossover

The multi-criterion FGGA utilizes a two-point crossover mechanism which works by crossing groups of chromosomes that lie between two randomly selected crossing sites. Intuitively, the crossover must be made more adaptive due to the fact that crossover operation must be more intense in the early than the last generations to ensure a balanced exploration and exploitation of the solution space. To achieve this, an adaptive crossover probability $p^c(t)$ at iteration (generation) t is formulated as follows:

$$p^c(t) = p_0^c + \frac{t}{T} \left(p_0^c - p_f^c \right) \tag{12.5}$$

where T is the defined maximum number of generations; p_0^m and p_f^m are the initial and final values of the crossover probability, respectively. The procedure for the adaptive crossover operator is presented in the following algorithm:

Algorithm 2 Adaptive crossover
  Input: p_0^m, p_f^m, population P
  Repeat

Step 1. Randomly select two parent chromosomes, P_1 and P_2, and select crossing sections for the two chromosomes.

Step 2. Compute the adaptive probability $p^c(t)$ at time t according to expression (12.5).

Step 3. With probability cross P_1 and P_2 by interchanging the crossing sections of the parent chromosomes.

Step 4. Obtain two offspring O_1 and O_2, which may have repeated items (called doubles) or missing items (called misses).

Step 5. Eliminate the doubles, using a constructive repair mechanism, avoiding the crossed items.

Step 6. Insert the missing items into groups, beginning from where doubles were eliminated, subject to group size limits.

Until (selection pool size is reached)
Return: Selection pool.

Figure 12.5 provides an example of the crossover mechanism based on two randomly selected parent chromosomes, $P1 = [1\ 2\ 3\ 4]$ with respective groups of items $\{4,7\}$, $\{1,3,5\}$, $\{2,6\}$, and $\{8,9\}$, and $P2 = [5\ 6\ 7\ 8]$ with corresponding groups of items $\{2,8\}$, $\{1,6\}$, $\{5,3\}$, and $\{7,4,9\}$.

According to the algorithm, the crossover mechanism is repeated until the desired number of new offspring is created. A new population, *newpop*, is formed by combining the best performing offspring and the parent population *oldpop*. The *newpop* is passed to the mutation operator.

12.4.6   Adaptive Mutation

FGGA uses the mutation operator to exploit the search space in the neighborhood of the current candidate solutions, a process known as exploitation or

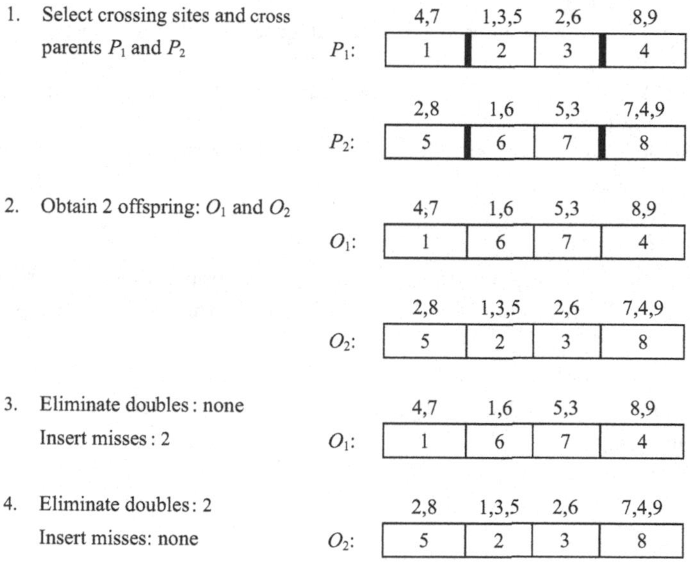

Fig. 12.5 An example of the adaptive crossover operation

intensification. To achieve this, selected chromosomes are subjected to slight per-turbations at a low probability p^m. Adaptive mutation assumes that mutation probability should be smaller in the early generations than in the latter generations, to allow convergence after sufficient exploitation of the solution space. The mutation probability $p^m(t)$ at generation t is determined by the following expression:

$$p^m(t) = p_0^m + \frac{t}{T}\left(p_f^m - p_0^m\right) \tag{12.6}$$

where T is the maximum number of generations; p_0^m and p_f^m are the initial and final values of the mutation probability, respectively.

For groups of fixed sizes such as the subcontractor selection example, the swap mutation mechanism is the most appropriate option. Swap mutation works exchanging items from selected groups of the chromosome to be mutated. If the operation results in an infeasible solution due to violation of hard constraints, a repair mechanism is utilized. The swap mutation procedure can be summarized into a four-stage algorithm as follows:

Algorithm 3 The adaptive mutation procedure

 Input: p_0^m, p_f^m, Population P;

 Compute new probability p_0^m

 For (*count* = 1 to p) **Do**   // p = population size

  Step 1. With probability p_0^m, select chromosome *count* for the mutation op-eration;

  Step 2. Randomly choose two different groups from the selected chromo-some;

  Step 3. Randomly select two items, one from each group of the selected groups;

  Step 4. Swap the selected items, and repair the resulting chromosome if necessary.

 End

 Return: P

Figure 12.6 further demonstrates the swap mutation mechanism, based on an example of a chromosome $P_1 = [1\ 7\ 3\ 4]$ with corresponding groups of items {1,3,7}, {6,5}, {2,9}, and {2,8}.

From this example, we see that two groups, that is, 7 and 4 are selected at random. Pairs of items from the two groups are then successively selected at random and swapped.

1. Select chromosome P1 for mutation, with probability p^m P_1:

1,3,7	6,5	2,9	2,8
1	7	3	4

2. Randomly select two groups: 7 and 4 P_1:

1,3,7	6,5	2,9	2,8
1	7	3	4

3. Randomly select items from the two groups: 5 and 2

1,3,7	6,5	2,9	2,8
1	7	3	4

4. Swap selected items: 2 and 5 Repair if need be P_1:

5,3	6,2	2,9	5,8
1	7	3	4

Fig. 12.6 An example of the adaptive mutation operation

12.4.7 Adaptive Two-Point Inversion

Inversion overturns items within two randomly selected inversion sites, howbeit, at a low inversion probability p^i. At each generation t, the inversion probability p^i is dynamically adjusted, so as to allow the population to converge to a common solution after a sufficient number of iterations. As such, a decay function is used to dynamically decrease the value of the probability:

$$p^i(t) = p_0^i e^{-\alpha(t/T)} \tag{12.5}$$

where t is the generation count; T is the maximum count; p_0 is the initial inversion probability; and $\alpha \in [0,1]$ is an adjustment factor which can also be varied according to the state of the search process. If the resulting chromosome violates any hard constraints, a repair mechanism is applied. The operator enables the crossover operator to involve a variety of groups in crossover operation. Groups or genes that are close together in the sequence are likely to be shifted together. The general procedure for the adaptive two-point inversion is shown in Algorithm 4.

Algorithm 4 Adaptive two-point inversion

Compute new value of $p^i(t)$ at iteration t;

Input: p_0^i, Population P;

For (*count* = 1 to p) **Do**   // p = population size
    Step 1. With probability $p^i(t)$, select chromosome *count* for the inversion operation;
    Step 2. Randomly select two inversion sites of the chromosome;
    Step 3. Overturn the selected groups that lie between the inversion sites;
    Step 4. Repair the chromosome if any hard constraints are violated;
End
Return: P

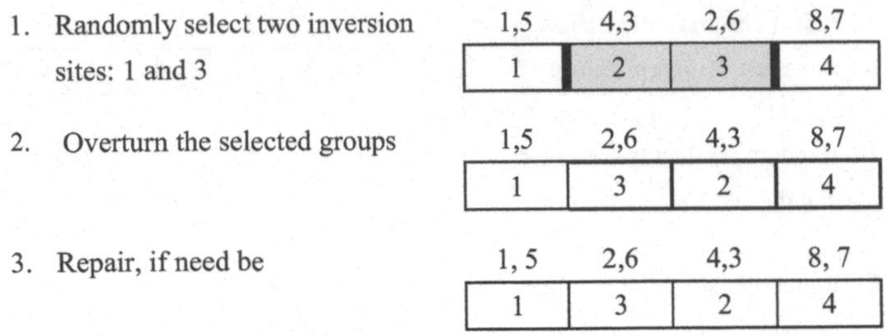

	1,5	4,3	2,6	8,7
1. Randomly select two inversion sites: 1 and 3	1	2	3	4

	1,5	2,6	4,3	8,7
2. Overturn the selected groups	1	3	2	4

	1, 5	2,6	4,3	8, 7
3. Repair, if need be	1	3	2	4

Fig. 12.7 Adaptive two-point inversion operator

Consider chromosome [1 2 3 4] with corresponding groups of items $\{1,5\}$, $\{4,3\}$, $\{2,6\}$, and $\{8,7\}$ as shown Fig. 12.7. Assume, that is, inversion sites 1 and 3 are selected. Inversion overturns (rewrites in the reverse order) groups 2 and 3 that lie between the two inversion sites, resulting in chromosome [1 3 2 4] with corresponding groups of items $\{1,5\}$, $\{2,6\}$, $\{4,3\}$, and $\{8,7\}$.

12.4.8 Replacement

At each iteration, the new population for the next generation $t + 1$ is obtained by replacing chromosomes in the old population t through the elitist strategy which ensures that best performing chromosomes in generation t are advanced into generation $t + 1$. As a rule of thumb, three best performing chromosomes are preserved and advanced into the next generation. To promote diversity, probabilistic crowding (Lianga and Leung 2011) should be applied to conserve and advance other potential chromosomes into the next generation.

12.4.9 Termination

The FGGA procedure continues until the termination condition is satisfied, that is, when (1) the maximum number of iterations, T, is reached, or (2) when there is no improvement in the current best solution within a prespecified number of iterations, or (3) when both 1 and 2 are satisfied.

12.5 Summary and Further Research

Supplier evaluation and selection are a complex problem characterized by fuzzy conflicting decision criteria, and imprecise management goals and aspirations. It was realized in this chapter that a number of conflicting criteria may have to be optimized simultaneously if the solution is to be satisfactory. Some of the criteria that were noted are price, lead time, quality, and number of suppliers or vendors selected. Based on an example of subcontractor selection, this chapter presented the supplier selection problem as a grouping problem where groups of tasks (items) can be assigned to each subcontractor, but at a specific cost. It was noted that in the subcontractor selection problem, the tasks may have due dates and precedence constraints, which make the problem even more complicated. This calls for advanced efficient, flexible, and interactive decision support systems that can handle fuzzy variables. To effectively address the fuzzy properties of the problem, a fuzzy multi-criterion grouping genetic algorithm was proposed to model the subcontractor selection. The algorithm uses a fuzzy evaluation approach to convert management goals and aspirations into normalized fuzzy membership functions.

Further applications of this model can be applied to similar supplier evaluation and selection problems characterized with (i) multiple suppliers and multiple commodities, (ii) multiple and often conflicting imprecise decision criteria, and (iii) due date restrictions, and/or precedence constraints.

References

Banaeian N, Mobli H, Fahimnia B, Nielsen IE, Omid M (2016) Green supplier selection using fuzzy group decision making methods: a case study from the agri-food industry. Comput Oper Res (In press). doi:10.1016/j.cor.2016.02.015

Chen CT, Lin CT, Huang SF (2006) A fuzzy approach for supplier evaluation and selection in supply chain management. Int J Prod Econ 102(2):289–301

Cheng M-Y, Tsai H-C, Sudjono E (2011) Evaluating subcontractor performance using evolutionary fuzzy hybrid neural network. Int J Project Manage 29:349–356

Choudhary D, Shankar R (2014) A goal programming model for joint decision making of inventory lot-size, supplier selection and carrier selection. Comput Ind Eng 71:1–9

Falkenauer E (1992) The grouping genetic algorithms-widening the scope of the GAs. JORBEL Belg J Oper Res Stat Comput Sci 33(1–2):79–102

Falkenauer E (1996) A hybrid grouping genetic algorithm for bin packing. J Heuristics 2(1):5–30

Galankashi MR, Helmi SA, Hashemzahi P (2016) Supplier selection in automobile industry: a mixed balanced scorecard–fuzzy AHP approach. Alexandria Eng J 55:93–100

Ghodsypour SH, O'Brien C (1998) A decision support system for supplier selection using an integrated analytic hierarchy process and linear programming. Int J Prod Econ 56–57:199–212

Hartmann A, Ling FYY, Tan JSH (2009) Relative importance of subcontractor selection criteria: evidence from Singapore. J Constr Eng Manag 135(9):826–832

Heidarzade A, Mahdavi I, Mahdavi-Amiri N (2016) Supplier selection using a clustering method based on a new distance for interval type-2 fuzzy sets: a case study. Appl Soft Comput 38 (2016):213–231

Jadidi O, Cavalieri S, Zolfaghari S (2015) An improved multi-choice goal programming approach for supplier selection problems. Appl Math Model 39(14):4213–4222

James T, Vroblefski M, Nottingham Q (2007) A hybrid grouping genetic algorithm for the registration area planning problem. Comput Commun 30(10):2180–2190

Kar AK (2015) A hybrid group decision support system for supplier selection using analytic hierarchy process, fuzzy set theory and neural network. J Comput Sci 6:23–33

Karsak EE, Dursun M (2015) An integrated fuzzy MCDM approach for supplier evaluation and selection. Comput Ind Eng 82:82–93

Kashan AH, Akbari AA, Ostadi B (2015) Grouping evolution strategies: an effective approach for grouping problems. Appl Math Model 39(9):2703–2720

Lianga Y, Leung K-S (2011) Genetic Algorithm with adaptive elitist-population strategies for multimodal function optimization. Appl Soft Comput 11:2017–2034

Mengshoel OJ, Goldberg DE (1999) Probability crowding: deterministic crowding with proba-bilistic replacement. In: Banzhaf W (ed) Proceedings of the international conference, GECCO-1999. Orlando, FL, pp 409–416

Mutingi M, Mbohwa C (2014) Home health care staff scheduling: effective grouping approaches. In: IAENG transactions on engineering sciences—special issue of the international multi-conference of engineers and computer scientists, IMECS 2013 and world congress on engineering, WCE 2013. CRC Press, Taylor & Francis Group, pp 215–224

Mutingi M, Mbohwa C (2016) Healthcare staff scheduling: emerging fuzzy optimization approaches. CRC Press, Taylor & Francis, New York

Polata G, Kaplan B, Bingol BN (2015) Subcontractor selection using genetic algorithm (Creative construction conference 2015, CCC2015). Proc Eng 123:432–440

Rajan JA, Ganesh K, Narayanan KV (2010) Application of integer linear programming model for vendor selection in a two stage supply chain. In: Proceedings of the 2010 international conference on industrial engineering and operations management, Dhaka, Bangladesh, 9–10 Jan 2010, pp 1–6

Rankovic V, Arsovski Z, Arsovski S, Kalinic Z, Milanovic I, Rejman-Petrovic D (2011) Multiobjective supplier selection using genetic algorithm: a comparison between weighted sum and SPEA methods. Int J Qual Res 5(4):289–295

Sodenkamp MA, Tavana M, Caprio DD (2016) Modeling synergies in multi-criteria supplier selection and order allocation: an application to commodity trading. Eur J Oper Res (in press). doi:10.1016/j.ejor.2016.04.015

Vijay W, Ravindran AR (2007) Vendor selection in outsourcing. Comput Oper Res 34(12):3725–3737

Weber CA, Current JR, Benton WC (1991) Vendor selection criteria and methods. Eur J Oper Res 50:2–18

Wright (1975) Consumer choice strategies/simplifying vs. optimizing. J Mark Res 12:60–67

Yahya S, Kingsman B (1999) Vendor rating for an entrepreneur development program: a case study using the analytic hierarchy process method. J Oper Res Soc 50:916–930

Yang PC, Wee HM, Pai S, Tseng YF (2011) Solving a stochastic demand multi-product supplier selection model with service level and budget constraints using genetic algorithm. Expert Syst Appl 38:14773–14777

Part IV
Conclusions and Extensions

Chapter 13
Further Research and Extensions

13.1 Introduction

Research presented in throughout this book has dealt with developments, advances, and innovations in grouping genetic algorithms (GGAs) and their variants such as grouping genetic algorithm (GGA), multi-criteria GGA, and fuzzy GGA (FGGA). These variants were equipped with new and improved genetic mechanisms and operators aimed at improving their overall optimization search efficiency. This endeavor was motivated by the realization of the ever-increasing complexity of grouping problems from various disciplines and by the upcoming new grouping problems in several industrial settings. Some of the most recent grouping problems realized include group-based estimation of discretionary accruals (Höglund 2013), reviewer group construction (Chen et al. 2011), team formation (Strnad and Guid 2010), modular design (Chen and Martinez 2012; Yu et al. 2011), group maintenance planning (Li et al. 2013; Do Van et al. 2013; Gunn and Diallo 2015; De Jonge et al. 2016), Wi-fi network deployment (Landa-Toress et al. 2013; Agustın-Blas et al. 2011), customer grouping (Ho et al. 2012), and home healthcare staff scheduling (Mutingi and Mbohwa 2016).

Though experimental tests and results based on various real-world benchmark problems have shown that grouping genetic algorithms are highly competitive in addressing these and other grouping problems, further extensions to the algorithms and their applications are quite possible and very relevant. Given the increasing complexity of the grouping problems, it is therefore useful to look into continued developments of the algorithm. In this view, this chapter seeks to highlight some of the possible extensions to the field of grouping genetic algorithms, from at least three perspectives, namely

1. Extensions to genetic mechanisms including the encoding scheme, fitness evaluation, selection, crossover, mutation, inversion, and termination,
2. Hybridization of GGA with other heuristic algorithms, which may lead to hyper-heuristics, and

© Springer International Publishing Switzerland 2017
M. Mutingi and C. Mbohwa, *Grouping Genetic Algorithms*,
Studies in Computational Intelligence 666,
DOI 10.1007/978-3-319-44394-2_13

3. Extensions to use of domain-specific heuristics that enhance the efficiency and effectiveness of algorithms.

Questions that arise at this point focus on the future: What does the future hold for grouping genetic algorithms? Looking at the possible future trends of industrial grouping problems, what are the likely challenges? How may the GGAs be improved in anticipation of the ever-increasing grouping problem domain, variety, size, and complexity? The main aim of this chapter is to suggest further possible extensions to the grouping genetic algorithms. Specific objectives and learning points in this regard are as follows:

1. To anticipate the possible growth trends and extension of the grouping problem domain, size, and complexity;
2. To evaluate the likely challenges that may arise from the likely growth of the problem domain and complexity; and
3. To visualize further extensions to the grouping genetic algorithm techniques that may further enhance the GGAs.

An accurate anticipation of the growth of the grouping problem domain and the associated complexity will enable developers of decision support systems to keep abreast with the future computational challenges. Appropriate advances to the algorithms are made possible.

The rest of this chapter is structured as follows: The next section presents further extensions to the possible application areas of the GGAs. Section 13.3 highlights future extensions to the techniques of the grouping algorithms that may be useful to further enhance the algorithms' efficiency and effectiveness. Section 13.4 provides concluding remarks. Finally, Sect. 13.5 summarizes and concludes this chapter.

13.2 Extension of the Application Domain

Having realized the capabilities and strengths of the proposed grouping algorithms, it is interesting to extend the horizons of the application domain of the variants of the algorithms. This is especially true, bearing in mind that, in the real world, the grouping problem domain, the problem size, and the associated complexity are ever-increasing. These problems arise from all industry disciplines. Extending the application domain of the GGAs is not only exciting, but also crucial in light of the growing problem domain. Extensions of the application of the grouping algorithms come in three different forms:

1. Application of the algorithms to new grouping problems, which usually demand new innovations in the algorithms;
2. Application of the algorithms to a more complex version of a previously solved problem; and

3. Application of a different variant of the grouping genetic algorithm on a known grouping problem.

Each of these forms usually comes with unique challenges that require a unique approach. Consider the first and second forms. When extending the application domain to a new problem or a more complex version of the previously solved problem, one possible and probably the easiest way is to begin from an analogous problem with similar characteristics.

By analyzing the characteristics of a new grouping problem to be solved, the structure of the algorithm can easily be adapted to the new problem. As an illustration, consider that the GGA approach has been applied to the problem of grouping learners for cooperative learning (Weitz and Jalassi 1992; Weitz and Lakshminarayanan 1996). Analogous to this scenario are other grouping problems such as crew grouping of soldiers (Liu et al. 2005), audit team construction (Dereli et al. 2007), registration area planning (Vroblefski and Brown 2006), and reviewers working in teams (Chen et al. 2011). By learning from the GGA variant that has been applied to grouping learners for cooperating learning, the GGA techniques can be extended or adapted to analogous problems with much ease. Table 13.1 lists variants of GGA and their respective applications presented earlier in this book,

Table 13.1 GGA variants, their applications, and the analogous grouping problems

Algorithm	Applications	Analogous grouping problem
GGA	Grouping learners for cooperative learning	Reviewers working in teams (Chen et al. 2011); audit team construction (Chen et al. 2011); research and development team construction
	Order batching in order picking warehouse systems	Classical vehicle routing problem (Toth and Vigo 2002); task assignment in home healthcare systems (Mutingi and Mbohwa 2014)
	Assembly line balancing	Workload allocation problem in a production line (Vidalis et al. 2005; Kalaycılar et al. 2016)
Multi-criteria GGA	Fleet size and mix vehicle routing problem	Home healthcare nurse scheduling problem (Mutingi and Mobhwa 2016); heterogeneous fixed fleet vehicle routing problem (Avci and Topaloglu 2016)
FGGA	Multi-criteria team formation	Cell formation problem (Onwubolu and Mutingi 2001); crew grouping of soldiers (Liu et al. 2005); audit team building (Dereli et al. 2007); reviewer group construction
	Multi-criteria examination timetabling	Capacitated exam timetabling (Burke et al. 2001); course timetabling Operating room scheduling; registration area planning (Vroblefski and Brown 2006)
	Modulardesign for sustainable manufacturing	Fuzzy group maintenance scheduling for sustainability (de Jonge et al. 2016; Gunn and Diallo 2015); design for disassembly/assembly

together with some other analogous problems; problems which have similar characteristics.

Variants of grouping genetic algorithms, including multi-criteria GGA, fuzzy GGA, and fuzzy multi-criteria GGA can be used to solve analogous grouping problems in the real world.

13.3   Further Extensions to Grouping Genetic Algorithms

It is evident that significant advances and extensions to GGAs were presented in this book. However, it is hoped that much more can still be done to investigate the performance of the variants of genetic operators, the hybridization of the grouping genetic algorithms with other heuristics, and the use of constructive domain-specific heuristics.

13.3.1   Variants of Grouping Genetic Operators

Variants of the algorithms can be obtained by incorporating variants of specific genetic mechanisms ranging from the encoding scheme, fitness evaluation, selection, crossover, mutation, inversion, and termination. Because there are various versions of each operator, there are several combinations of these variants that may be incorporated into a GGA variant. For instance, the mutation operator has been known to have several, including split mutation, merge mutation, and swap mutation. It may be worthwhile experimenting on the effect of using each one of the variants on the overall computational efficiency and effectiveness. Similarly, investigations on the effects of variants of other operators can be carried out. Furthermore, other genetic mechanisms and procedures can be investigated. For instance, the effect of applying different group encoding schemes, different evaluation methods, self-adaptive methodologies, and replacement strategies can be investigated. The assumption is that these mechanisms and operators have different impacts on the overall effectiveness and efficiency of the ultimate algorithm. Therefore, it will be quite interesting to investigate further on the resulting effect of their interactions on the overall performance of the grouping genetic algorithms.

13.3.2   Hybridizing GGA with Heuristic Algorithms

Knowing that there are other competitive global optimization algorithms apart from genetic algorithms, developers of grouping genetic algorithms should consider borrowing efficient optimization techniques from other algorithms. The end goal is to improve the efficiency of the algorithm. In addition, hybrid grouping algorithms can be developed by infusing at least one operator from other efficient evolutionary algorithms. For example, efficient operators can be obtained from fast evolutionary approaches such as particle swarm optimization, ant systems, memetic algorithms, hybrid genetic algorithms (Torabi et al. 2006), as well as other local search approaches such as tabu search and simulated annealing. Noteworthily, the goal is to improve the efficiency and effectiveness of the grouping algorithm and not to make the resulting hybrid algorithm more cumbersome and computationally slow. Techniques and heuristics can also be borrowed from other competing grouping evolution strategies (Kashan et al. 2015).

13.3.3   Further Use of Domain-Specific Heuristics

Domain-specific heuristics are constructive heuristics that take advantage of the specific structures of the problem under study. More specifically, constructive heuristics utilize specific knowledge about the grouping structure, the objective function(s), and the associated constraints (especially the hard constraints). With prior knowledge of the problem, expert systems or intelligence can be built into specific mechanisms of the algorithms, from the group encoding scheme down to termination.

In some cases, group encoding can be designed to take care of possible infeasibilities that may arise during the optimization process. By exploiting the nature of the hard constraints at hand, the GGA designer can incorporate some of the hard constraints into the group encoding scheme. Furthermore, the designer should always make it a point to structure the encoding scheme so that it is easier to decode the information and perform fitness evaluation without computational difficulties or delays. Other genetic mechanisms may need to be taken into account when developing a coding scheme, so as to come up with an efficient and effective coding scheme, which improves the efficiency of the overall algorithm.

Oftentimes, it is important to develop an efficient initialization algorithm. Such an algorithm is usually made up of domain-specific heuristics, basic heuristics, greedy heuristics, or seed algorithms that generate good starting solutions for the grouping genetic algorithm. However, it is important to note that the knowledge of the problem structure, specifically the grouping structure, is a prerequisite to the development of good domain-specific heuristics.

A careful design of the main genetic operators (crossover, mutation, and inversion) is of paramount importance if the grouping algorithm is to produce optimal or near-optimal solutions efficiently. In the case that violation of hard constraints by genetic operators is inevitable, repair mechanisms should be put in place to curb the infeasibilities. However, repair mechanisms are usually very problem-specific and should be developed based on sufficient prior knowledge of the problem structure. It is crystal clear at this point that the GGA designer needs to have in-depth knowledge of the problem structure in order to develop efficient domain-specific constructive heuristics that will enhance the grouping algorithm from the group encoding down to the termination point.

13.4 Concluding Remarks

Enhanced grouping genetic algorithms work extremely well, and in most cases significantly better than other competitive algorithms, insofar as grouping problems are concerned. This was evidenced by several illustrative computational experiments on problems from various disciplines. The performance of the algorithms was comparatively superior to the conventional genetic algorithm, other heuristics, and other evolutionary algorithms. As such, grouping genetic algorithms are appreciably effective and efficient when applied to complex real-world grouping (or clustering, or partitioning) problems. However, the actual implementation of these algorithms to specific grouping problems does not go without its own challenges.

The first and foremost challenge is to visualize the grouping structure of the problem at hand. When the problem structure is mastered well, the next challenge is how to exploit the structure so as to come up with a suitable group encoding scheme. The mechanisms of the grouping genetic operators that follow need to be structured with the encoding scheme in mind. Other accessories such as initialization, evaluation, and replacement strategies have to be structured according to the encoding scheme as well. As discussed in several chapters of this book, offline fine-tuning is essential for genetic parameters such as crossover, mutation, and inversion probabilities. Much more than that, self-adaptive and dynamic mechanisms have to be built into the genetic operators to further enhance the effectiveness and efficiency of the algorithms, especially when addressing complex grouping problems. Over and above all, the overall interaction of all the genetic operators and the related mechanisms should be harmonized with the encoding scheme. Thus, unlike the old assertion that genetic algorithms are characterized with strong performance without problem-dependent information, grouping genetic algorithms encourage reliance on exploitation of the grouping structure of the problem being solved. Moreover, the proposed grouping genetic algorithms in this book encourage hybridization of the algorithms with domain-specific constructive heuristics.

In view of the above issues and challenges, it is strongly believed that the future of grouping genetic algorithms lies in developing spirited efforts on the following:

1. Strong exploitation of the grouping structure of the problem in order to develop the most appropriate encoding scheme;
2. Enhancement of computational efficiency through advanced fine-tuning, self-adaptive and dynamic mechanisms in genetic operators;
3. Use of fuzzy theoretic and fuzzy logic techniques to enhance the genetic operator mechanisms and to model fuzziness in grouping problems;
4. Use of domain-specific constructive heuristics to augment the global optimization process; and
5. Hybridization of the grouping algorithms with other competitive heuristic methods

With much rigor on the above, addressing the ever-growing complexities in real-world grouping problems is highly possible and can be much easier than ever expected.

13.5 Summary

Throughout this book, advances and developments of variants of grouping genetic algorithms (GGAs) have been presented, tested, and applied to various grouping problems in a wide range of industrial disciplines. In spite of these revelations, it is hoped that further extensions to the algorithms are quite possible, and this could lead to further fruitful applications. It is therefore useful to look into the possible developments of the algorithm and their applications. In this view, this chapter highlighted some of the possible further extensions to GGA, from three perspectives, namely (i) extensions to genetic mechanisms and grouping techniques such as the encoding scheme, fitness evaluation, selection, crossover, mutation, inversion, and termination, (ii) hybridization of GGA with other heuristic algorithms, which may lead to hyper-heuristics, and (iii) extensions to use of domain-specific heuristics that enhance the efficiency and effectiveness of algorithms. Following these GGA extensions, potential application areas of the techniques were suggested and presented. Over and above the suggested extensions, there still remains more potency in the use of fuzzy theoretic concepts and fuzzy logic control both in modeling fuzzy problem variables and guiding the global optimization process.

Reflecting on the research work presented in this book, it is with much hope that the advances, extensions, and applications in the work will be a significant addition to the body of knowledge in expert systems, computational algorithms, operations research, operations management, and intelligent systems. GGAs are quite handy, when applied to medium- to large-scale real-world grouping problems in various industrial disciplines, including healthcare, information and communication technology, tertiary education, maintenance planning, warehouse and logistics planning, economics and business, production and manufacturing systems, and design. The knowledge will therefore go a long way toward equipping the computational scientists, operations analysts, industrial engineers, and other decision makers.

References

Agustın-Blas LE, Salcedo-Sanz S, Vidales P, Urueta G, Portilla-Figueras JA (2011) Near optimal citywide WiFi network deployment using a hybrid grouping genetic algorithm. Expert Syst Appl 38(8):9543–9556

Avci M, Topaloglu S (2016) A hybrid metaheuristic algorithm for heterogeneous vehicle routing problem with simultaneous pickup and delivery. Expert Syst Appl 53:160–171

Burke EK, Bykov Y, Petrovic S (2001) A multi-criteria approach to examination timetabling. In: Burke EK, Erben W (eds) Practice and theory of automated timetabling: selected papers from the 3rd international conference. Lecture Notes in Computer Science 2079, pp 118–131

Chen AL, Martinez DH (2012) A heuristic method based on genetic algorithm for the baseline-product design. Expert Syst Appl 39(5):5829–5837

Chen Y, Fan Z-P, Ma J, Zeng S (2011) A hybrid grouping genetic algorithm for reviewer group construction problem. Expert Syst Appl 38:2401–2411 de Jonge B, Klingenberg W, Teunter R, Tinga T (2016) Reducing costs by clustering maintenance activities for multiple critical units. Reliab Eng Syst Saf 145:93–103

Dereli T, Baykasoglu A, Das GS (2007) Fuzzy quality-team formation for value added auditing: a case study. J Eng Tech Manage 24(4):366–394

Gunn EA, Diallo C (2015) Optimal opportunistic indirect grouping of preventive replacements in multicomponent systems. Comput Ind Eng 90:281–291

Ho GTS, Ip WH, Lee CKM, Mou WL (2012) Customer grouping for better resources allocation using GA based clustering technique. Expert Syst Appl 39:1979–1987

Höglund H (2013) Estimating discretionary accruals using a grouping genetic algorithm. Expert Syst Appl 40:2366–2372

Kalaycılar EG, Azizoğlu M, Yeralan S (2016) A disassembly line balancing problem with fixed number of workstations. Eur J Oper Res 249(2):592–604

Kashan AH, Akbari AA, Ostadi B (2015) Grouping evolution strategies: an effective approach for grouping problems. Appl Math Model 39(9):2703–2720

Landa-Torres I, Gil-Lopez S, Del Ser J, Salcedo-Sanz S, Manjarres D, Portilla-Figueras JA (2013) Efficient citywide planning of open WiFi access networks using novel grouping har-mony search heuristics. Eng Appl Artif Intell 26:1124–1130

Li F, Ma L, Sun Yong, Mathew J (2013) Group maintenance scheduling: a case study for a pipeline network. In: Engineering Asset Management 2011. Proceedings of the sixth annual world congress on engineering asset management (Lecture Notes in Mechanical Engineering), Springer, Duke Energy Center, Ohio, pp 163–177

Liu H, Xu Z, Abraham A (2005) Hybrid fuzzy-genetic algorithm approach for crew grouping. In: Proceedings of the 5th international conference on intelligent systems design and applications (ISDA'05), pp 332–337

Mutingi M, Mbohwa C (2014) A Fuzzy-based particle swarm optimization approach for task assignment in home healthcare. S Afr J Ind Eng 25(3):84–95

Mutingi M, Mbohwa C (2016) Fuzzy grouping genetic algorithm for homecare staff scheduling. In: Mutingi M and Mbohwa C (ed) Healthcare staff scheduling: emerging fuzzy optimization approaches, 1st edn. CRC Press, Taylor & Francis, New York, pp 119–136

Onwubolu GC, Mutingi M (2001) A genetic algorithm approach to cellular manufacturing systems. Comput Ind Eng 39:125–144

Strnad D, Guid N (2010) A Fuzzy-Genetic decision support system for project team formation. Appl Soft Comput 10(4):1178–1187

Torabi SA, Fatemi Ghomi SMT, Karimi B (2006) A hybrid genetic algorithm for the finite horizon economic lot and delivery scheduling in supply chains. Eur J Oper Res 173:173–189

Toth P, Vigo D (2002) The vehicle routing problem. SIAM Monograph on Discrete Mathematics and Applications, Philadelphia

Van Do P, Barros A, Bérenguer C, Bouvard K, Brissaud F (2013) Dynamic grouping maintenance with time limited opportunities. Reliab Eng Syst Saf 120:51–59

Vidalis MI, Papadopoulosb CT, Heavey C (2005) On the workload and phaseload allocation problems of short reliable production lines with finite buffers. Comput Ind Eng 48(4):825–837

Vroblefski M, Brown EC (2006) A grouping genetic algorithm for registration area planning. Omega 34:220–230

Weitz RR, Jelassi MT (1992) Assigning students to groups: a multi-criteria decision support system approach. Decis Sci 23:746–757

Weitz RR, Lakshminarayanan S (1996) On a heuristic for the final exam scheduling problem. J Oper Res Soc 47:599–600

Yu S, Yang Q, Tao J, Tian X, Yin F (2011) Product modular design incorporating life cycle issues —group genetic algorithm (GGA) based method. J Clean Prod 19(9–10):1016–1032

Index

© Springer International Publishing Switzerland 2017
M. Mutingi and C. Mbohwa, *Grouping Genetic Algorithms*,
Studies in Computational Intelligence 666,
DOI 10.1007/978-3-319-44394-2

Printed in the United States
By Bookmasters